CYTOKINES

THE MEDICAL PERSPECTIVES SERIES

Editors:

Andrew P. Read *Department of Medical Genetics, University of Manchester, St Mary's Hospital, Hathersage Road, Manchester M13 0JH, U.K.*

Terrence Brown *Department of Biochemistry and Applied Molecular Biology, UMIST, Manchester M60 1QD, U.K.*

Oncogenes and Tumor Suppressor Genes

Cytokines

Autoimmunity (due early 1992)

Genetic Engineering (due early 1992)

The Human Genome (due early 1992)

CYTOKINES

M.J Clemens
*Department of Cellular and Molecular Sciences,
Division of Biochemistry, St George's Hospital Medical School,
Cranmer Terrace, London SW17 0RE, U.K.*

© BIOS Scientific Publishers Limited, 1991

All rights reserved. No part of this book may be reproduced or transmitted, in any form or by any means, without permission.

First published in the United Kingdom 1991 by
BIOS Scientific Publishers Limited,
St Thomas House, Becket Street, Oxford OX1 1SJ.

A CIP catalogue record for this book is available from the British Library.

ISBN 1 872 748 70 8

For Jenny

Typeset by Enset Photosetting, Midsomer Norton, U.K.
Printed by Information Press Ltd, Oxford, U.K.

PREFACE

This book aims to describe the current state of knowledge concerning the regulatory proteins known as cytokines, which have a very wide range of important physiological functions in the human body. There has been an enormous explosion of knowledge concerning these proteins in recent years, a trend which shows every likelihood of continuing. This is the result of several developments in the fields of cell and molecular biology, embryology and clinical medicine. In particular, the immense power of recombinant DNA techniques has enabled molecular biologists to isolate and sequence the genes and messenger RNAs that code for many of the cytokines and their receptors. At the same time, developments in immunology have led to the production of monoclonal antibodies that are specific for different cytokines and these are proving to be powerful tools for analysis of the functions of the many members of this group.

Whilst the techniques of basic science have provided the means by which our knowledge has progressed, the impetus for the application of these approaches has been the realization that the cytokines play fundamental roles in the control of many physiological functions in the body and are implicated in a large variety of disease states. Important examples of diseases in which cytokines are undoubtedly involved include many different cancers, autoimmune disorders, virus infections (including AIDS), and inflammatory diseases. It is therefore important that we should understand as much as possible about the structure and function of the cytokines, their roles in regulating normal processes, and the ways in which their regulatory activities are disrupted in disease. In addition, there is the very practical consideration that it may be possible to use cytokines therapeutically to control many of these diseases, once we know how they work *in vivo*. This book sets out to explain these various aspects of cytokine biology – what cytokines are, what they do, how they work at the cellular level, their roles in health and disease, and how they are being used in clinical medicine today. Each chapter ends with a limited selection of references where the reader will find further details of the topics covered. In choosing these I have concentrated largely on recent reviews, together with a few research papers that indicate current exciting developments in the relevant fields.

Since progress in all these areas will undoubtedly be rapid the book ends with a look into the future in an attempt to predict further applications in a field where the combination of basic science and clinical medicine should yield beneficial results.

M. J. Clemens

ACKNOWLEDGMENTS

I wish to thank all members of my research group for putting up with my many hours of absence from the laboratory during the preparation of this book. I am also extremely grateful to Barbara Bashford for all the artwork for the book.

My own research in the area of interferon action and related fields has been funded by research grants from the Cancer Research Campaign, the Leukaemia Research Fund, the Gunnar Nilsson Cancer Research Trust, the Wellcome Trust and the Medical Research Council of Great Britain. I thank all these bodies for their financial support.

CONTENTS

Abbreviations   ix

1 Introduction to cytokines   1
What are cytokines?   1
Classes of cytokines   3
References   18
Further reading   19

2 Cytokine genes and proteins   21
Structure and expression of cytokine genes   21
Cytokine production   25
How are cytokines assayed?   30
References   31
Further reading   32

3 How do cytokines work?   33
Receptors for cytokines   33
Signal transduction pathways   38
Regulation of gene expression   48
Control of cell surface proteins   51
References   53
Further reading   54

4 Biological roles of cytokines   57
Cytokine networks   57
Regulation of cell growth and differentiation   62
Cytokines and the immune system   68
References   72
Further reading   72

5 Cytokines in health and disease   75
Homeostatic regulation   75

Wound healing	79
The immune system	80
Control of virus infections	82
Cytokines and cancer	85
Control of inflammation	92
References	95
Further reading	96

6 Therapeutic uses of cytokines — 99

Control of normal processes	99
Regulation of immune function	102
Infectious diseases	103
Cytokines in cancer therapy	105
Inflammation	109
Future prospects for clinical applications of cytokines	109
References	110
Further reading	111

Appendix A. Glossary — 113

Index — 117

ABBREVIATIONS

AIDS	acquired immunodeficiency syndrome
BFU-E	burst-forming unit – erythroid
BMP	bone morphogenetic protein
CFU	colony forming unit
CFU-E	colony forming unit – erythroid
CML	chronic myelogenous leukemia
CSF	colony stimulating factor
DAG	diacylglycerol
dsRNA	double-stranded RNA
EGF	epidermal growth factor
EPO	erythropoietin
FGF	fibroblast growth factor
G protein	GTP-binding protein
GM-CSF	granulocyte/macrophage colony stimulating factor
HIV	human immunodeficiency virus
IFN	interferon
IgE	immunoglobulin E
IGF-I	insulin-like growth factor I
IL	interleukin
IP_3	inositol 1,4,5-trisphosphate
LAK	lymphokine-activated killer
LIF	leukemia inhibitory factor
LPS	lipopolysaccharide
MCP-1	monocyte chemoattractant protein-1
MHC	major histocompatibility complex
NGF	nerve growth factor
PAF	platelet-activating factor
PDGF	platelet-derived growth factor
PGI_2	prostacyclin
PK	protein kinase
$PtdIns(4,5)P_2$	phosphatidylinositol 4,5-bisphosphate
rhEPO	recombinant human erythropoeitin
SLE	systemic lupus erythematosus
TGFα/β	transforming growth factor α/β
TNF	tumor necrosis factor
TS	tumor suppressor

1
INTRODUCTION TO CYTOKINES

1.1 What are cytokines?

The term 'cytokine' is applied to proteins produced by cells in response to a variety of inducing stimuli; they are secreted by their producer cells and then influence the behavior of target cells. Classical polypeptide hormones also fulfil this definition, but by convention are not classed as cytokines since they are produced by specific endocrine organs (e.g. the pituitary gland), whereas a cytokine may be produced by more than one cell type and in a number of tissues. In many respects, however, their actions are similar: both cytokines and classical polypeptide hormones act after secretion from producer cells and must bind to specific receptors on the surfaces of target cells. The resulting effects inside the target cell are brought about by signal transduction across the plasma membrane. The basic elements of this model are illustrated in *Figure 1.1*.

The nature of the target cell for a particular cytokine is determined by the presence of specific receptors. These may be present on the surface of the same cell as produces

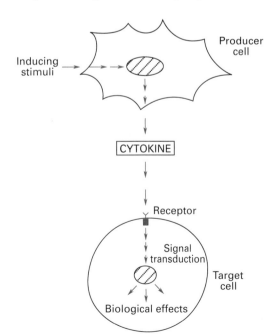

Figure 1.1: Essential elements in the mode of action of cytokines. Cytokines are secreted (usually as a result of new synthesis) by specific cell types in response to inducing stimuli. They bind to specific receptors on the surfaces of target cells, eliciting biochemical changes that result in signal transduction. In most cases signals are sent to the nucleus and cause changes in gene expression.

the cytokine, in which case autocrine effects (regulation of cellular activity by its own product) are possible. Alternatively, the cytokine may work only on other cell types which are not themselves producers. If these target cells lie close by the producer cell, the term 'paracrine regulation' is used to describe the process. The target cells may also occur in distant parts of the body, or in other organs and tissues, giving rise to a type of regulation analogous to the mode of action of polypeptide hormones. *Figure 1.2* summarizes these three types of cytokine–cell interaction.

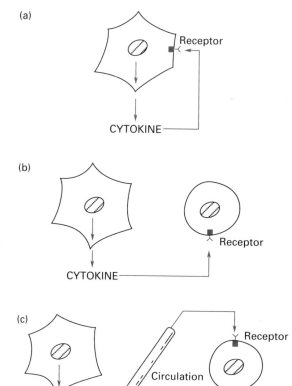

Figure 1.2: *Three types of cytokine–cell interaction. Only cells possessing suitable receptors can respond to a cytokine.* **(a)** *Autocrine stimulation occurs when a cell possesses receptors to respond to cytokines it produces itself.* **(b)** *A localized paracrine stimulation occurs if a different cell type, lying near the producer cell, has the appropriate receptors.* **(c)** *Cytokines may also be released into the bloodstream or other body fluids and interact with target cells elsewhere in the body (similar to endocrine stimulation performed by classical hormones).*

Although cytokines are a very diverse group of molecules, with a variety of effects, most share certain structural and functional features. They are all proteins that are mostly rather small (typically 15–30 kd). Some are modified before secretion by addition of carbohydrate side-chains and are thus glycoproteins. Some cytokines are synthesized by their producer cells as larger precursors that are later cleaved to give the biologically active molecule. As expected of agents that have a number of regulatory roles, cytokines are rarely produced at a constant rate but rather are induced (or suppressed) by specific stimuli to which the body needs to respond. Moreover, their lifetime in the bloodstream or other extracellular fluids into which they are secreted is usually short, ensuring that they act for only a limited period. Clearance and destruction of cytokine molecules is probably a regulated process, just as synthesis and secretion of these proteins is tightly controlled under normal circumstances.

Comparisons of various cytokines in humans and other animals show that there has been much conservation of structure and function during their evolution. This suggests that the need for cytokines arose rather early during the evolution of multicellular organisms and that the structures of many regions of these proteins are critical for function. Most of the physiological roles of the cytokines are indeed related specifically to the requirements of multicellular organisms, and are concerned either with co-ordination of processes between different cell types or with the response to environmental stresses (Table 1.1). Thus we find that cytokines have essential roles in control of cell proliferation and differentiation during embryonic development and in later life. They are involved in regulating the immune response to foreign antigens and invading organisms, and they are important in mounting other forms of defense against infection by viruses and other pathogens. Essential processes, such as cellular renewal and wound healing, the development of cellular and humoral immunity and inflammatory responses, all require participation of a range of cytokines. It is not surprising therefore that many diseases involving disruption of these processes are associated with altered regulation of cytokine production and action. Knowledge about the roles of cytokines in these situations is beginning to open the way to their therapeutic use against some forms of cancer, serious virus infections and inflammatory diseases, as will be described later in this book.

Table 1.1: Physiological roles of cytokines in multicellular organisms

---

Control of cell proliferation
Control of cell differentiation and phenotype
Regulation of hematopoiesis
Regulation of immune responses
Control of host defenses against viral and parasitic infections
Regulation of inflammatory responses and fever
Control of cytotoxic and phagocytic cells
Wound healing
Tissue remodelling and bone formation
Influences on cellular metabolism and control of nitrogen balance

---

1.2 Classes of cytokines

Identification and characterization of the many cytokines now known has resulted in a variety of names and classifications. In this chapter the major groups of cytokines are described according to current nomenclature. It must be remembered, however, that this classification owes as much to historical accident as to functional similarities between members of the different groups. Conversely, there is considerable overlap in the biological effects and mechanisms of action of cytokines currently placed in different categories.

1.2.1 Growth factors

It has been known for many years that a number of small polypeptides are able to promote growth and division of various cell types in tissue culture. These molecules are known collectively as growth factors *(Table 1.2)*. Many of these are able to act on a variety of target cells and their names may not accurately indicate their specificity. Moreover, the term 'growth factor' may be misleading: in many cases cell differentiation rather than proliferation is the main consequence of treating a cell type with a growth factor.

Epidermal growth factor (EGF) [1] is so named because of its ability to stimulate proliferation of basal epithelial cells of the skin. Human EGF is a 53-amino-acid polypeptide that is synthesized (as a much larger precursor) by a range of normal cell types and is found in almost all body fluids. It exerts a mitogenic effect (promotion of cell proliferation) on several types of cell, including corneal epithelium and tracheal tissue, but it can also stimulate bone resorption. These properties have led to the idea that EGF plays an important role in tissue remodelling during development and in wound repair throughout life.

Table 1.2: *Properties of the major growth factors in man*

Growth factor	Number of amino acids	Molecular mass of native factor (daltons)	Principal sources
EGF	53	c. 6000	Many cell types
PDGF	125 (A chain) 160 (B chain)	28–35 000	Megakaryocytes; monocytes; endothelium; smooth muscle
Acidic FGF	140	16–19 000	Neuroectoderm
Basic FGF	146	16–19 000	Many cell types
IGF-I	70	c. 7000	Liver; neural tissues
IGF-II	67	c. 7000	Liver
NGF	118 (B chain)	c. 130 000 (polymer)	Salivary gland; neural tissue

Platelet-derived growth factor (PDGF) [2] is produced not only by platelets but also by endothelial cells, activated macrophages and several other cell types. There is in fact a family of three PDGF molecules comprising dimers of two sub-units (A and B) in all possible combinations. Both A and B chains are synthesized as larger precursors which are shortened after dimerization. The AA form of PDGF has less mitogenic potency than the AB and BB forms, but the significance of this is not clear. PDGFs are believed to act in a paracrine manner on connective tissue cells and may be involved in mediating chronic inflammatory processes as well as the hyperplasia associated with tissue repair and wound healing. Although PDGF alone can sometimes drive cells through DNA synthesis and cell division, the mitogenic action of PDGFs appears to be due to their ability to commit cells to entering the cell cycle in response to other growth factors. Some evidence suggests that PDGF-like proteins play a role in development of the central nervous system during embryogenesis, but further work is needed to clarify the roles of the various PDGF isoforms in different tissues.

The two structurally related factors known as acidic and basic fibroblast growth factors (FGFs) [3] are products of distinct genes and consist of single chains of 140 and 146 amino acids respectively. They are mainly produced by cells of the pituitary gland and parts of the brain, although other cell types also synthesize FGFs. Both molecules are potent growth stimulatory factors for neuroectodermal cells, vascular endothelial cells, fibroblasts and other cells. FGFs have angiogenic (blood capillary promoting) activity and they may play a role in tissue vascularization, as yet another influence on wound healing and tissue regeneration. FGFs show amino acid sequence relationships with the products of some proto-oncogenes (see Chapter 5). In normal development, expression of these proto-oncogenes is confined to embryonic tissues, suggesting that the main role of such FGF-like proteins is to regulate tissue growth and organization during early fetal life.

Although insulin is not a cytokine (see Section 1.1), it must be included in the group of mitogenic growth factors because it can promote DNA synthesis and cell proliferation in a wide range of cells in culture. It probably does so not only by binding to its own receptor (which has structural and functional similarities with the EGF receptor), but also by binding to a receptor for another factor, insulin-like growth factor I (IGF-I) [4]. IGF-I (formerly called somatomedin C) is a true cytokine which bears a structural resemblance to insulin but is produced by many mammalian cell types. It probably mediates the mitogenic activity of pituitary growth hormone, since it is produced in response to growth hormone stimulation. There is another insulin-like growth factor, IGF-II [4], which is mitogenic (it was originally called 'multiplication stimulating activity') and has structural resemblances to insulin and IGF-I. Unlike insulin, IGF-I and IGF-II are single chain proteins of 70 and 67 amino acids respectively, both synthesized as larger precursor molecules. Insulin and the IGFs each have distinct receptors but each has some ability to bind to each other's receptor (except that insulin does not bind to the IGF-II receptor). *In vivo,* IGFs are found in plasma bound to carrier proteins, a form in which they may be inactive. IGF-I is undoubtedly involved in growth promotion *in vivo* and mediates the growth-hormone dependent growth of skeletal cartilage and mammary gland. IGF-II may play a similar role in fetal development. The growth-inducing activity of other hormones may also be due to their induction of IGFs; for example, estrogens stimulate IGF-I production in the uterus. It is less clear whether insulin itself is important as a growth factor in the intact organism (as opposed to cells in culture). However, its well-documented metabolic actions, such as stimulation of glucose uptake and its ability to stimulate protein synthesis, are probably permissive for cell growth. Possibly insulin is necessary but not sufficient for mitogenic activity *in vivo,* although the structure of its receptor and the signal transduction pathway employed (see Chapter 3) may argue for some direct mitogenic role.

Nerve growth factor (NGF) [5] is a cytokine that has very different effects depending on the target cell. It is probably more important as a neurotropic factor, necessary for differentiation and survival of neural tissues, than as a mitogen. Nevertheless, receptors for NGF are present on other cell types and the factor can accelerate wound healing and exert effects on mast cells (basophils) and platelets. Its major sources are submaxillary and prostate glands and neural tissues.

This list of mitogenic growth factors is not exhaustive: others will be discussed in the following sections, and there are numerous other less well-studied factors which may be equally important *in vivo* (*Table 1.3*). No doubt others are still waiting to be discovered.

1.2.2 Lymphokines

Cytokines that are produced by or act upon cells of the immune system are known as lymphokines. This is in many ways an artificial distinction since, as will be seen,

Table 1.3: Recently described growth factors

Growth factor	Molecular mass of native factor (daltons)	Principal sources	Biological effects
Neuroleukin	c. 56 000	Muscle; brain; kidney	Involved in motor neuron development
Amphiregulin	c. 9000	Breast tumor cells	Similar to EGF/TGFα (on only some cell types)
Hepatocyte growth factor	c. 90 000 (dimer)	Liver; spleen	Regulation of liver regeneration following hepatic injury

lymphokines have much in common with other types of cytokine, in their production, in the way they act and in the nature of their target cells. Factors commonly referred to as lymphokines include interferon γ and the interleukins. Interleukins act as intercellular signalling agents between lymphocytes. Interleukins 1–8 are described below and summarized in *Table 1.4*; one or two new interleukins are added to the list each year.

The generic name 'interleukin' was proposed in 1979 for factors produced and released by activated T lymphocytes that act on other lymphocytes to produce biological effects. In fact, the group of interleukins also includes species produced primarily by cells of other hematopoietic lineages, such as monocytes. Interleukin 1 (IL-1) is an example of such a 'monokine' [6]. It is produced by monocytes in response to stimulation by fever-inducing agents such as bacterial lipopolysaccharide (LPS), and was previously known as 'lymphocyte activating factor' or 'endogenous pyrogen' because of its fever-inducing properties. Two forms of IL-1, IL-1α and IL-1β, were cloned and sequenced in the mid-1980s. As with several growth factors described earlier in this chapter, IL-1α and β are both synthesized as larger precursor proteins from which the active molecules are derived by proteolytic cleavage. The mature proteins contain 159 and 153 amino acids, respectively. Perhaps surprisingly, the two forms of IL-1 are very similar in their actions on target cells despite having only 25% amino acid sequence homology. Probably both IL-1 species evolved from a common ancestral gene, but the advantages to the organism of having two distinct forms are not obvious.

Although monocytes are a major source of IL-1, almost every cell type in the body can produce IL-1 under appropriate conditions. There are probably almost as many stimulating agents for IL-1 production as there are producer cell types; it can even stimulate its own synthesis, giving rise to an interesting case of positive feedback regulation. Expression of the two genes for IL-1α and β, and the release of these proteins from cells, are also differentially regulated.

As with all polypeptide cytokines and hormones, there are specific receptors for the interleukins on the surface of target cells. IL-1α and β share a common receptor despite the structural differences between the two molecules, which explains the similarity in their biological effects. It appears that two forms of IL-1 receptor, of high and low affinity, are found on a wide variety of cell types, potentially giving IL-1 an almost universal role in regulating cellular activities. The lower-affinity receptor is an 80 kd surface glycoprotein that belongs to the immunoglobulin superfamily (see Chapter 3). As expected of a cytokine that can act on a range of cell types, the effects of binding

Table 1.4: Properties of the interleukins

Interleukin	Number of amino acids	Molecular mass of native factor (daltons)	Principal sources
IL-1α	159	c. 17 500	Monocytes and many other cell types
IL-1β	153	c. 17 300	Monocytes and many other cell types
IL-2	133	c. 15 400	T lymphocytes
IL-3 (multi-CSF)	133	15–25 000	T lymphocytes
IL-4	129	15–19 000	T lymphocytes
IL-5	125	45–60 000 (dimer/trimer)	T lymphocytes
IL-6	184	c. 21 000	T lymphocytes; monocytes and many other cell types
IL-7	152	20–28 000	Bone marrow stromal cells
IL-8	72	c. 8000	Monocytes and macrophages

IL-1α and β to their receptor vary according to the target. In general, IL-1 mediates two classes of effect, stimulation of immune responses (including induction of synthesis of other lymphokines), and production of inflammatory responses via IL-1-induced production of prostaglandins (see Chapter 3).

Interleukin 2 (IL-2) [7] was formerly known as T-cell growth factor, indicating its importance in controlling proliferation of T lymphocytes. It is produced almost exclusively by T lymphocytes themselves in response to antigenic stimulation, an effect that can be mimicked *in vitro* by treating the cells with plant lectins such as phytohemagglutinin. Again, it is synthesized as a precursor, although in this case only an N-terminal 20-amino-acid signal sequence is removed to produce the mature protein of 133 amino acids. IL-2 is variably glycosylated on threonine residue 3 but this does not appear to be necessary for its activity.

The cell surface receptor for IL-2 consists of two polypeptides, a 55 kd α sub-unit (known as the Tac antigen) and a 70–75 kd β sub-unit. The β sub-unit converts the receptor from a low to a high affinity form, and it is mainly through this high affinity receptor that IL-2 exerts its biological effects. An interesting mechanism for regulating cellular sensitivity to IL-2 involves inducible expression of the α sub-unit following T lymphocyte stimulation. The larger (β) sub-unit appears already to be present on the surfaces of unstimulated cells. IL-2 and its receptor constitute an important example of autocrine regulation of cellular functions, since the multifarious responses of T lymphocytes to antigenic or mitogenic stimulation include early induction of both IL-2 and the Tac antigen. Simultaneous production of both the interleukin and the high affinity form of its receptor allows self-stimulation of T cells, which is largely responsible for the subsequent entry of the cells into a state of active proliferation. There is a role for IL-1 in this process since it can enhance production of IL-2. IL-2 receptors are also found on other cell types within the immune system, including natural killer cells, monocytes/macrophages and B lymphocytes. Thus IL-2 (as well as IL-1 and other interleukins) probably also regulates the activity of these cells in various ways. The complex interactions between the interleukins (and other cytokines) in controlling cellular activities, and particularly their role in the immune response, are considered in more detail in Chapter 4.

Interleukin 3 (IL-3) [8], like IL-2, is exclusively produced by T lymphocytes, but it is not strictly an interleukin since its targets include a range of non-lymphocyte cell types within several hematopoietic lineages. Mature human IL-3 consists of 133 amino acids; the actual molecular mass of the protein is somewhat variable, since it can be heavily glycosylated at several possible sites, although with little consequence for its biological activity. There is little sequence similarity with IL-2, despite the similar number of amino acids.

The IL-3 receptor is only partially characterized but is believed to be a 140 kd phosphoprotein. Stimulation of this receptor by IL-3 results in proliferative responses in cells of both the erythroid and myeloid lineages. In this respect the biological activities of IL-3 and granulocyte/macrophage colony stimulating factor (see Section 1.2.3) overlap, although IL-3 may have a broader specificity. IL-3 can promote proliferation of precursors of erythroid cells, megakaryocytes, macrophages, neutrophils, eosinophils and mast cells, which explains why it had so many different names before it was definitively characterized by cloning and sequencing. Possibly this apparently wide spectrum of cell targets means that IL-3 acts on early multipotential cells (but not primitive stem cells); however, the physiological relevance of this remains unclear since IL-3 is produced by peripherally activated T lymphocytes but these targets lie within bone marrow. It has been speculated that IL-3 may augment the activities of the more lineage-restricted colony stimulating factors in response to physiological stresses such as acute infections. A role for IL-3 in allergic responses may also be postulated since it stimulates production of histamine-containing mast cells. As with most other interleukins, there are probably a number of networks by which IL-3 modulates the effects of other cytokines, for example, by inducing expression of receptors.

Interleukin 4 (IL-4) [9] is another example of a cytokine exclusively produced by T lymphocytes but acting on a range of target cells. Its principal role *in vivo* is probably to promote proliferation of B lymphocytes which have been activated by antigen binding or other cell-surface mediated events. This activity of IL-4 is reflected in its earlier names, B-cell growth factor-I or B-cell stimulating factor-1. Mature human IL-4 contains 129 amino acids, with two possible sites for protein glycosylation. As with other cytokines, glycosylation does not appear to be necessary for biological activity. IL-4, granulocyte/macrophage colony stimulating factor and interferon γ share a similar pattern of gene organization and some amino acid sequence similarities, perhaps indicating some distant evolutionary relationship.

Although B lymphocytes are a major cell target for IL-4 *in vivo*, the 60 kd receptor for this lymphokine is also present on T lymphocytes, macrophages, mast cells and other cells of the hematopoietic system, and possibly also on non-hematopoietic cell types. Expression of the IL-4 receptor on B and T lymphocytes can be up-regulated by IL-4 itself – another example of positive feedback control during cellular stimulation. Besides stimulating proliferation of B cells, IL-4 can enhance the production of immunoglobulin E, implicating it in the development of allergic responses. There are a number of interesting relationships between the effects of IL-4 and those of other lymphokines. For example, IL-4 may interfere with the stimulatory effects of IL-2 on lymphokine-activated killer cells (see Chapter 4), and interferon γ can in turn inhibit the responses of B lymphocytes to IL-4.

Interleukin 5 (IL-5) [10], like many of the other lymphokines described here, was originally identified by virtue of several different effects, for example, on B-lymphocyte activation (in the mouse) and eosinophil differentiation. Earlier names for it included B-cell growth factor-II and eosinophil-differentiating factor. IL-5 is produced by activated T lymphocytes and is secreted as a 125-amino acid glycosylated protein.

The IL-5 receptor is not yet well characterized, but chemical crosslinking of IL-5 to its putative receptor produces a 92.5 kd complex. In human eosinophil precursor cells the principal effect of binding IL-5 is to stimulate cell proliferation and maturation. This specifically enhances production of eosinophils, with little effect on other hematopoietic cell lineages. IL-5 may also enhance the maintenance and cellular activity of mature eosinophils, which is probably crucial for antibody-dependent control of parasitic and other infections by these cells. In some allergic diseases there is overproduction and activation of eosinophils, suggesting that IL-5 may be important in regulating the progress of inflammatory conditions.

The identification of interleukin 6 (IL-6) [11] provides a fascinating historical insight into how molecular biology can cause apparently separate fields of research to converge. Until IL-6 was cloned and sequenced its many biological activities had led to it being defined as an interferon (IFNβ2); a factor required for differentiation and antibody secretion by B lymphocytes (B-cell stimulatory factor-2); a growth factor for B-cell tumors (plasmacytoma/hybridoma growth factor); and an agent that stimulates production of acute-phase proteins by liver (hepatocyte stimulating factor). Here then is yet another example of a lymphokine produced by T lymphocytes that acts on a bewildering variety of target cells. In fact, IL-6 is also synthesized by monocytes/macrophages, fibroblasts and endothelial cells, making it, together with IL-1, one of the most ubiquitously active cytokines. Furthermore, a large array of seemingly unrelated inducers, including viruses, bacterial LPS, IL-1 and phorbol esters, can lead to IL-6 synthesis and secretion.

The mature form of human IL-6 is a 184-amino acid protein with two possible glycosylation sites. Comparisons with the granulocyte colony stimulating factor gene and protein suggest that both are derived from a common ancestor. The widely distributed IL-6 receptor appears to be a single-chain molecule of 467 amino acids that, like the IL-1 receptor, is a member of the immunoglobulin superfamily. Expression of this receptor can be regulated so that, for example, activation of B lymphocytes leads to induction of the receptor and hence proliferating and differentiating B cells can respond to IL-6. The biological activities of IL-6 are so varied that it is difficult to predict which are of greatest physiological importance. Very probably the ability to promote differentiation of B lymphocytes into antibody-secreting plasma cells, and the induction of acute-phase protein synthesis in the liver in response to injury or inflammation are crucial contributions of IL-6 to the normal host response to environmental stresses. In addition, IL-6 can act as an enhancing signal for various T lymphocyte activities (IL-2 production and cell proliferation) and for IL-3-dependent development of hematopoietic precursors in the bone marrow. IL-6 can also induce a febrile response *in vivo*.

Just as lymphokines like IL-5 and IL-6 appear to regulate the activity of terminally differentiated target cells, there are other cytokines that control proliferation of earlier progenitor cells. An example is interleukin 7 (IL-7). This factor, produced by bone marrow stromal cells, can regulate the growth of B-lymphocyte precursors (pro-B cells) that have not yet undergone immunoglobulin gene rearrangement. Human IL-7 is a potentially glycosylated protein of 152 amino acids, for which a cDNA has recently been described [12]. Little is yet known about its receptor, but it may not be restricted to pro-B cells since mature T lymphocytes can also apparently respond to this lymphokine by enhanced production of IL-2 and the IL-2 receptor. This is perhaps surprising, given the origin of IL-7 in bone marrow stroma, but it indicates that IL-7, IL-1 and IL-6 may have some overlapping activity.

As indicated earlier, the list of cytokines now defined as interleukins continues to grow as further cDNAs are cloned and sequenced. The name interleukin 8 (IL-8) has been assigned to a macrophage-derived factor that is chemotactic for neutrophils [13]. Mature IL-8 contains only 72 amino acids and is probably not glycosylated. Amino acid sequence analysis indicates that it may be a member of a family of inflammatory or growth-regulatory proteins. One role of IL-8 and related factors is probably to attract neutrophils to sites of tissue damage or infection, leading to increased phagocytic and cytolytic activity. Lymphocytes, fibroblasts, endothelial cells and also macrophages can probably produce chemotactic proteins like IL-8, and it is likely that other cytokines such as IL-1 and tumor necrosis factor can stimulate them to do so. IL-8 can also attract and activate T lymphocytes. Recently, interleukins 9, 10 and 11 have also been characterized and sequenced [14–16]. The full range of functions that can be ascribed to these new factors remains to be established.

1.2.3 Colony stimulating factors

The term 'colony stimulating factor' relates to the way in which growth and differentiation of specific lineages of blood cells have been assayed in the past, using growth in semi-solid agar or other tissue culture systems. Certain cytokines promote the appearance of 'colonies' of cells *in vitro,* leading to the name now in common usage. Although only an operational definition, the name colony stimulating factor (CSF) is conceptually useful since this group of cytokines function mainly to regulate proliferation and differentiation of various types of hematopoietic cell (e.g. granulocytes, macrophages and erythroid cells). There is functional overlap between CSFs and some of the interleukins described earlier. For example, IL-3 was formerly called eosinophil-CSF and multi-CSF, amongst other names, because it promotes growth and differentiation of both myeloid and erythroid cell precursors. Thus its activities overlap those of granulocyte/macrophage-CSF (see below). Similarly, IL-5 is a lineage-specific factor for proliferation and differentiation of eosinophils, as described earlier.

In order to appreciate the roles of the CSFs (summarized in *Table 1.5*) it is necessary to understand how hematopoietic cell lineages are thought to arise from a small population of multipotential stem cells in the bone marrow (*Figure 1.3*). These stem cells undergo continual self-renewal, providing a pool of cells that can become committed to any of the hematopoietic lineages. Such committed cells must still proliferate, expanding the number of progenitor cells, before eventually undergoing terminal differentiation to produce the various mature blood cells. The latter no longer proliferate under normal circumstances; thus their numbers are determined by the earlier growth and differentiation, and it is at these levels that the CSFs exert their effects *in vivo*.

As its name indicates, granulocyte/macrophage-CSF (GM-CSF) [17–19] supports proliferation of bone-marrow derived progenitors of both granulocytes and macrophages. It is now believed to act on multipotential cells at a somewhat later stage of differentiation than those cells which are a target for IL-3. It also exerts effects on mature granulocytes and macrophages. Proliferation of other hematopoietic lineages, such as erythroid and megakaryocyte precursors, may also be sensitive to GM-CSF, although IL-3 is probably more important for these cell types.

Human GM-CSF is a 127-amino acid single polypeptide chain with several potential glycosylation sites (*Table 1.5*). Surprisingly, given how its effects overlap those of IL-3, there is no homology between the primary sequences of the two cytokines. However, their genes lie close together on human chromosome 5. The receptor for GM-CSF may be quite large (180 kd), although smaller forms also exist, and there may be another receptor that also has affinity for IL-3. A receptor of the latter type would explain the common effects of GM-CSF and IL-3.

GM-CSF is produced by activated T lymphocytes and also by other cells including fibroblasts, endothelial cells and monocytes. Production of GM-CSF by stromal cells in the bone marrow regulates the development of early hematopoietic progenitor cells by a paracrine mechanism. Synthesis and secretion of GM-CSF are themselves regulated, for example, by IL-1 or interferon γ, and often occur in response to antigenic stimulation. Thus GM-CSF probably has a major role in host defense mechanisms requiring increased numbers of granulocytes and macrophages and this could be more important than the part it plays in normal hematopoiesis. The pleiotropic actions of GM-CSF on mature myeloid cell types mean that it can also activate the very cells whose proliferation it has induced – for example, to increase cytocidal or oxidative activity against invading microbes. The phagocytic activity of neutrophils towards bacteria is also stimulated by GM-CSF.

Like GM-CSF, granulocyte (G)-CSF [17–19] acts on both granulocyte precursors and the mature cells of this hematopoietic lineage, but its effects are largely restricted to this lineage alone. It is a protein of 174 amino acids and like all the secreted CSFs is derived from a larger precursor (*Table 1.5*). There are probably carbohydrate side-chains covalently linked to the backbone of the secreted protein. As expected, the G-CSF receptor is found on granulocytes (especially neutrophils). It has a molecular

Table 1.5: Properties of the colony stimulating factors

Factor	Number of *amino acids	Molecular mass of native factor (daltons)	Principal sources
Multi-CSF (IL-3; see *Table 1.4*)			
GM-CSF	127	18–22 000	T lymphocytes and many other cell types
G-CSF	174	c. 19 600	Monocytes; fibroblasts; endothelium
M-CSF	149–224	40–45 000 (dimer)	Monocytes; fibroblasts; endothelium
EPO	166	34–39 000	Kidney and some other cell types
LIF	179	c. 38 000	T lymphocytes; carcinoma cells

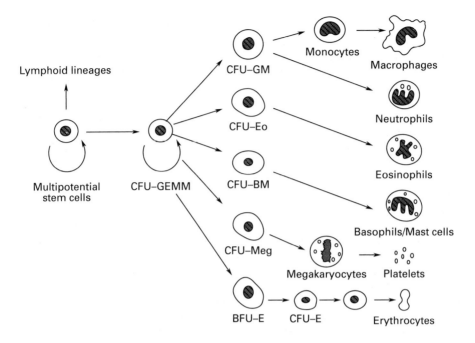

Figure 1.3: Origin of hematopoietic cell lineages in the bone marrow. All blood cell types are derived by differentiation from self-renewing multipotential stem cells in the bone marrow. These give rise to proliferating CFU–GEMM (colony forming units–granulocyte erythrocyte monocyte megakaryocyte). CFU–GEMM can become committed to differentiation in any of the lineages shown, under the influence of various cytokines, to form CFU or BFU (burst-forming units). Macrophages and neutrophils are derived from the same progenitor CFU–GM (granulocyte/monocyte) population. Cell proliferation ceases as CFU undergo terminal differentiation.

mass of 150 kd and is distinct from other cytokine receptors also found on these cell types, such as those for GM-CSF or IL-3.

Cell types able to produce G-CSF are generally similar to those that secrete GM-CSF, except for T lymphocytes. As with GM-CSF, expression of G-CSF occurs predominantly in response to specific stimuli, and bacterial LPS (endotoxin) is a potent inducer. IL-1 and mitogens also stimulate G-CSF production. These observations again suggest that G-CSF plays a major role in the fight against bacterial and other infections. As well as enhancing proliferation of granulocyte precursors in bone marrow, G-CSF activates neutrophils in several ways, stimulating phagocytosis, superoxide production (see Chapter 4) and antibody-mediated cytotoxicity against targets such as tumor cells. G-CSF also apparently has chemotactic properties, inducing neutrophils to migrate towards sites of inflammation. As with GM-CSF, we see a single factor which can both promote production of a particular cell type and activate this cell type.

Just as granulocytes have a specific CSF to regulate their activities, so macrophages are controlled by a relatively specific cytokine, macrophage (M)-CSF [17–19]. However, the physiological role of M-CSF (also formerly called CSF-1) is less clear than for the CSFs described above. Monocytes and macrophages are the main sources of M-CSF in normal tissues. Its structure is complicated as there appear to be polypeptide chains ranging in size from 149 to 224 amino acids and with variable degrees of glycosylation. Alternative processing of the M-CSF mRNA can generate different lengths of coding sequence (see Chapter 2), and the biologically active soluble forms of M-CSF exist as dimers of these variously sized sub-units.

The M-CSF receptor is well characterized, in contrast to the other CSF receptors, because it has been shown to be the product of the *FMS* proto-oncogene (see Chapter 5). This receptor, like many others, is a large protein (about 150 kd) that spans the plasma membrane. Like the receptors for EGF, insulin and PDGF, it has tyrosine kinase activity, and indeed it shares some structural features with the PDGF receptor. The significance of tyrosine kinase activity for signal transduction will be discussed in Chapter 3.

M-CSF probably functions in an autocrine manner *in vivo*, since monocytes and macrophages both produce this cytokine and carry the receptor for it. On stimulation by M-CSF, macrophages secrete IL-1 and prostaglandin E, and show enhanced synthesis of peroxide, at least in the mouse. Presumably human M-CSF fulfils a similar role in promoting these host defense mechanisms. Activation of phagocytosis and cell killing have also been noted in response to M-CSF. Thus this cytokine is one more of the weapons with which the body attempts to fight off invading organisms or abnormal cells (e.g. tumors). There are also indications that M-CSF has a role in controlling development, not only of macrophage progenitor cells but also of the placenta. In the latter case it appears to be uterine epithelial cells which produce M-CSF in response to estrogens.

A survey of the hematopoietic growth factors would not be complete without consideration of erythropoietin (EPO) [19]. This factor is distinct from the CSFs in several ways. It is synthesized primarily by kidney glomerular cells in response to the level of oxygenated erythrocytes and it qualifies more as a hormone than a cytokine. Nevertheless, the function of EPO is to stimulate proliferation and differentiation of erythroid precursor cells in bone marrow.

Although the existence of EPO has been known for many years, only with the cloning of the cDNA was its structure defined. Mature EPO is a heavily glycosylated protein of 166 amino acids. Its receptor is found only on the surface of erythroid precursor cells and has a molecular mass of nearly 100 kd, almost half contributed by carbohydrate side-chains. Possibly this molecule interacts with another protein sub-unit to mediate cellular responses to EPO.

Like some of the CSFs discussed earlier, EPO is active only on cells that are already committed to their particular lineage. It can directly stimulate proliferation of a cell type called burst-forming unit–erythroid (BFU-E), giving rise to colonies. At a later stage of differentiation these cells generate colony forming units (CFU-E) that are also sensitive to EPO, producing erythroblasts that synthesize hemoglobin. The effects of EPO on these cellular targets can be synergistic with those of lymphokines such as IL-1 and IL-3. Yet again we have a factor that is responsible for the generation of differentiated cells of a specific lineage and for stimulation of their differentiated functions. As a result EPO is essential for generation of normal levels of erythrocytes and for the adaptation of these levels to changing physiological requirements in man. It may also fulfil an additional function in maintaining platelet levels in the blood, since EPO can stimulate the growth of megakaryocyte colonies.

This section has described the major cytokines involved in the regulation of hematopoietic cell growth and differentiation. It must be stressed, however, that the sophisticated control mechanisms involved in these processes require complex interactions between multiple factors. Notably, the effects of various CSFs are strongly modulated by other cytokines already mentioned (e.g. IL-1 and IL-6). In addition to these, another factor has recently been described – leukemia inhibitory factor (LIF) [20]. This 179-amino acid glycosylated polypeptide can induce myeloid cell differentiation but is also able to promote proliferation and inhibit differentiation of embryonic stem cells. It is not clear what role LIF might play in the regulation of hematopoiesis *in vivo*. Interactions between cytokines during the growth and differentiation of bone marrow precursors of blood cells are considered in more detail in Chapter 4.

1.2.4 Transforming growth factors

Two cytokines which promote proliferation of certain cells in culture have been named transforming growth factor (TGF) α and β *(Table 1.6)*. However, it was an historical accident that linked these two factors together, and the names do not imply any structural or functional relationship. Indeed, their true physiological roles may be diametrically opposed since TGFβ is a growth *inhibitor* rather than a mitogen for most cell types.

TGFα [21] is a polypeptide of only 50 amino acids, derived from a much larger precursor molecule. It has structural features in common with EGF (see Section 1.2.1). Consistent with this structural similarity, TGFα binds to the same cell surface receptor as EGF and can presumably bring about the same cellular effects. There appears to be no other receptor for TGFα, which creates an exceptional situation in which two growth factors bind to only one receptor species on their target cells. Surprisingly, the same receptor is exploited by a related growth factor encoded by vaccinia virus, perhaps explaining why this virus causes local benign hyperplasia in infected skin lesions.

The main physiological role of TGFα may be to stimulate growth of tissues such as placenta and kidney during embryonic development. It could perhaps be the fetal equivalent of EGF, although why both factors are required remains unclear. Consistent with its role in promoting cell proliferation in a variety of tissues, TGFα can be abnormally produced by some tumors, leading to uncontrolled cell division. TGFα also appears to be potentially important for resorption and formation of bone, perhaps accounting for the ability of tumors to destroy bone in some cancer patients. Like EGF, TGFα could also play a role in wound healing since it can be produced by both skin cells and macrophages.

TGFβ [22, 23] actually consists of a small family of structurally related polypeptides. The functional factor is a dimer (usually a homodimer) of sub-units containing 112 amino acids each. There are at least three possible sub-unit sequences, designated β1, β2 and β3, derived from much larger precursors by proteolysis. The three sequences are highly homologous, suggesting strong evolutionary pressure to conserve critical structural features in the dimeric molecule. Many cell types can produce TGFβs and this, together with the wide range of effects observed, suggests a fundamental physiological role for these molecules. Blood platelets are a particularly good source.

It is not clear whether different forms of TGFβ must share the same receptor. However, there is a large (about 300 kd) receptor to which TGFβ1, β2 and β3 can all bind. It comprises a polypeptide core (approximately 30% of the mass) to which large oligosaccharide and glycosaminoglycan side-chains are covalently linked. This unusual structure raises intriguing questions about the biochemical mechanisms by which the TGFβ family work.

The biological effects of the TGFβ group are numerous and diverse. Although TGFβ was identified as a growth-stimulating agent, and can act as such on some mesenchymal

Table 1.6: Properties of the transforming growth factors

Factor	Number of amino acids	Molecular mass of native factor (daltons)	Principal sources
TGFα	50	c. 6000	Tumor cells; keratinocytes; macrophages
TGFβ1 TGFβ2 TGFβ3	112	c. 25 000 (dimer)	Megakaryocytes; macrophages; lymphocytes; bone

cells, this effect is atypical for most target cells. It probably requires the autocrine production and action of the BB form of PDGF (see Section 1.2.1) and could be most important during embryogenesis. The more characteristic response of cells to TGFβ is growth inhibition. Indeed, TGFβ can antagonize the effect of positive growth factors on fibroblasts, epithelial cells, muscle cells and hematopoietic cells. It seems possible that TGFβ can exert autocrine growth inhibition, which might be important in preventing uncontrolled proliferation of normal cells in response to mitogenic growth factors such as EGF or the FGFs. Failure of such a negative regulatory system could contribute to the development of tumors, as discussed in Chapter 5. Along with other cytokines discussed earlier, TGFβ probably plays significant roles in tissue repair and regeneration. It can accelerate wound healing, partly by promoting production of connective tissue components such as collagen and fibronectin, and partly by attracting other cell types such as monocytes to sites of wound repair by chemotaxis. The TGFβs, or related molecules such as the bone morphogenetic proteins, also promote cartilage formation and bone remodelling.

Yet other inhibitory roles for the TGFβs are in control of hematopoiesis and lymphocyte activation. For example, TGFβ treatment blocks the mitogenic effect of IL-2 on T lymphocytes and can block antibody production by B lymphocytes. Macrophage activation and inflammatory responses are also inhibited by TGFβ, which may limit the extent of inflammation under normal circumstances.

The TGFβ group is a subfamily of a broader class of growth and differentiation-regulating factors (see Chapter 2). This includes not only the bone morphogenetic proteins mentioned above but also the inhibins and activins, and Müllerian inhibitory substance. These proteins play various roles in differentiation of endocrine cell types in pituitary, ovarian and testicular tissues.

1.2.5 Tumor necrosis factors

As with many cytokines, the name 'tumor necrosis factors' (TNFs) reflects the history of these agents, and is misleading because although TNFs can induce regression of some tumors, it gives no clue to their many other actions [24]. There are two species of TNF, α and β *(Table 1.7)*, that show considerable homology in their amino acid sequences. Human TNFα is a 157-amino acid protein derived from a larger precursor; it exists as a complex of three such polypeptide chains. It is produced mainly by macrophages in response to activation by agents such as bacterial LPS (endotoxin). TNFα has also been called 'cachectin' because of its role in inducing cachexia (tissue wasting, negative nitrogen balance and loss of body weight). TNFβ (formerly called lymphotoxin) is larger (171 amino acids), and is produced mainly by activated T lymphocytes. It is glycosylated *in vivo*, unlike TNFα. Despite these differences, both species of TNF share a widely distributed common 70 kd receptor. This is probably constitutively expressed by most cells, but its level can be regulated; for example, dramatic induction of the TNF receptor occurs upon activation of T lymphocytes, and up-regulation also occurs in cells treated with interferons. Such modulation of TNF receptor levels provides potential mechanisms for altering cellular sensitivity to these cytokines, and with T cells the induction by antigens or mitogens of both TNFβ and its receptor can permit autocrine regulation of cellular activities. This is reminiscent of the situation with IL-2 and its receptor in the same cell type. In contrast, bacterial endotoxin induces TNFα but down-regulates its receptor in macrophages.

Table 1.7: Properties of the tumor necrosis factors

Factor	Number of amino acids	Molecular mass of native factor (daltons)	Principal sources
TNFα	157	c. 51 000 (trimer)	Macrophages
TNFβ	171	60–70 000 (trimer)	T lymphocytes

Other cell types can also produce TNFα (e.g. T cells and fibroblasts) and perhaps TNFβ (B lymphocytes). Additional inducers include other cytokines, especially IL-1 and IL-2. In general, production of the TNFs forms part of the response to invasion by micro-organisms and is an important part of host defenses against infection. In conjunction with other factors such as IL-2, TNFs can stimulate proliferation of both T and B cells, and will enhance antibody secretion, resembling the effects of IL-1 on lymphocyte activity. Cell–cell recognition may also be facilitated by TNFs since they stimulate expression of cell surface major histocompatibility antigens; another facet of this may be activation of cytolytic and cytotoxic cells.

In contrast to the positive actions of TNFs on lymphocytes, they generally inhibit proliferation of hematopoietic cells (except possibly granulocyte precursors). TNFs may also function in the monocyte/macrophage lineage to induce differentiation and production of other cytokines. Again, there may be an autocrine loop for control of macrophage activity by a product of this cell type. It is therefore likely that TNFs play an important role in control of inflammatory responses. Activity of granulocytes and eosinophils is enhanced by TNFs, leading to increased phagocytosis and other mechanisms of cell killing. These effects undoubtedly contribute to the anti-tumor actions of the TNFs *in vivo* (see Chapter 5) but, in addition, TNFs have a direct cytostatic or cytotoxic effect on some tumor cells, demonstrable *in vitro*. Sometimes this anti-proliferative action is due to stimulation of cell differentiation but other, more lethal, mechanisms may be involved, such as generation of arachidonic acid metabolites and free radicals.

1.2.6 Interferons

Interferons (IFNs) were originally identified as agents produced by virus-infected cells which can protect cells against further viral infections (hence the name, derived from viral 'interference'). It is now clear, however, that IFNs can elicit many other changes in cellular behavior, including effects on cell growth and differentiation and modulation of the immune system [25]. The multiple species of IFNs are now classified into three groups, α, β and γ (*Table 1.8*) on the basis of their distinct antigenic properties. An earlier classification defined them as either Type I, acid-stable IFNs (the α and β species) or Type II, acid-labile IFN (now called IFNγ); IFNs were also called leukocyte, fibroblast and immune IFNs, approximately corresponding to the modern α, β and γ designations. Now that we can clone and sequence the IFNs, classifications based on cell of origin or physicochemical properties have become redundant.

Table 1.8: Properties of the interferons

Interferon	Number of amino acids	Molecular mass of native IFN (daltons)	Principal sources
IFNα s[a]	166–172	16–27 000	T lymphocytes; B lymphocytes; monocytes; fibroblasts
IFNβ	166	c. 20 000	Fibroblasts
IFNγ	143	c. 50 000 (dimer)	T lymphocytes; natural killer cells

[a] The IFNαs constitute a large family of closely related proteins.

In humans the IFNα subtype encompasses a surprisingly large number of molecular species. At least 16 different IFN genes code for proteins of 166–172 amino acids, some of which may be modified by O-linked glycosylation. These IFN sequences are all closely related and this, together with highly conserved intrachain disulfide bridges, presumably accounts for their common antigenic properties. In contrast to the diversity of the IFNα group, there is only one IFNβ species in man, also containing 166 amino acids. It has less homology with the α IFN sequences and is an N-linked glycoprotein. Note that at one time the cytokine now designated as IL-6 was called IFNβ2; however, the amino acid sequence of IL-6 shows only a distant evolutionary relationship with IFNβ, and IL-6 has only a weak (if any) ability to induce an anti-viral state in its target cells. As with IFNβ, there is only one human IFNγ. The mature form is a glycosylated polypeptide of 143 amino acids, forming a dimer in its native state. IFNγ bears only slight structural similarities to α IFNs or to IFNβ but does have true anti-viral activity.

All the α IFNs and IFNβ appear to share a common receptor, a 130 kd glycoprotein that is widely distributed on different cell types. The existence of only a single receptor for multiple species of IFNs raises the question, why have so many distinct molecular forms of IFNα evolved? Even more puzzling, the ratios of different biological activities elicited by the α IFNs can vary between different molecular species which apparently all act through the same receptor. Perhaps the affinity of an IFN for the receptor somehow determines the nature of the cellular response; this would imply a rather complex mechanism of interaction of the IFNs with their receptor, possibly involving more than one site. Alternatively, there may be a family of structurally similar IFNα receptors, with distinct effects on the cell.

Consistent with its lack of structural homology with the other IFNs, IFNγ possesses a distinct receptor, widely distributed on different cell types. Almost half of its 90 kd molecular mass is contributed by extensive carbohydrate side-chains on the extracellular part of the molecule.

As might be predicted for such a large family of cytokines with almost ubiquitously distributed receptors, IFNs have very varied physiological roles. Production of IFNα or β is induced by viral and other forms of infection, or the presence of foreign cell types and antigens. It is not clear in every case what specific molecules are responsible for induction, but double-stranded RNA and cytokines can be good inducers. With IFNγ, T-lymphocyte activating agents and foreign antigens are inducers. There is much overlap between different cell types in both the inducers and the species of IFN induced. The major producer cell types are: for IFNα – lymphocytes, monocytes and macrophages; for IFNβ – fibroblasts and some epithelial cells; for IFNγ – activated T lymphocytes.

In addition to the 'classical' anti-viral activities that all the IFNs elicit in their target cells (see Chapter 5), the biological consequences of an IFN binding to its receptor can include inhibition of cell proliferation, induction of cell differentiation and changes in cell morphology. A possible physiological role for such effects may be to limit proliferation of normal cells, perhaps by autocrine or paracrine feedback, countering the actions of positive growth factors on cell division. Such a role might require a low but constitutive level of IFN production in the absence of any external inducing agent. Enhancement of histocompatibility antigen expression on many cells and stimulation of immunoglobulin–Fc receptor expression on macrophages are also observed following IFN treatment. Fc receptor expression is required for antibody-dependent cellular cytotoxicity of macrophages. B lymphocytes can be induced to increase antibody production by low concentrations of IFNα or β. Moreover, in contrast to the generally inhibitory effects of IFNs on cell proliferation, B cells may actually show enhanced growth in response to IFNγ. An additional effect of the α and β IFNs is activation of natural killer cells that may be responsible for destruction of virus-infected cells or tumor cells *in vivo*. These cells are also activated by IL-2. Overall, it would appear that IFNs are of great importance as part of the body's defenses against foreign organisms and abnormal cell types, and that a multiplicity of mechanisms has evolved by which these defenses are mobilized.

References

1. Carpenter, G. and Cohen, S. (1990) *J. Biol. Chem.*, **265**, 7709.
2. Ross, R., Bowen-Pope, D.F. and Raines, E.W. (1990) *Phil. Trans. R. Soc. Lond. (Biol.)*, **327**, 155.
3. Burgess, W.H. and Maciag, T. (1989) *Ann. Rev. Biochem.*, **58**, 575.
4. Humbel, R.E. (1990) *Eur. J. Biochem.*, **190**, 445.
5. Black, I.B., DiCicco-Bloom, E. and Dreyfus, C.F. (1990) *Curr. Top. Dev. Biol.*, **24**, 161.
6. Platanias, L.C. and Vogelzang, N.J. (1990) *Am. J. Med.*, **89**, 621.
7. Kuziel, W.A. and Greene, W.C. (1990) *J. Invest. Dermatol.*, **94**, 27S.
8. Metcalf, D. (1989) in *Peptide Regulatory Factors*. Edward Arnold, London, p. 17.
9. Jansen, J.H., Fibbe, W.E., Willemze, R. and Kluin-Nelemans, J.C. (1990) *Blut*, **60**, 269.
10. Sanderson, C.J. (1990) *Int. J. Cell Cloning*, **8**, 147.
11. Kishimoto, T. (1989) *Blood*, **74**, 1.

12. Namen, A.E., Williams, D.E. and Goodwin, R.G. (1990) *Prog. Clin. Biol. Res.* **338,** 65.
13. Matsushima, K., Morishita, K., Yoshimura, T. *et al.* (1988) *J. Exp. Med.,* **167,** 1883.
14. Donahue, R.E., Yang, Y.-C. and Clark, S.C. (1990) *Blood,* **75,** 2271.
15. Moore, K.W., Vieira, P., Fiorentino, D.F., Trounstine, M.L., Khan, T.A. and Mosmann, T.R. (1990) *Science,* **248,** 1230.
16. Paul, S.R., Bennett, F., Calvetti, J.A. *et al.* (1990) *Proc. Natl. Acad. Sci. USA,* **87,** 7512.
17. Whetton, A.D. (1990) *Trends Pharmacol. Sci.,* **11,** 285.
18. Nicola, N.A. (1989) *Ann. Rev. Biochem.,* **58,** 45.
19. Dexter, T.M., Garland, J.M. and Testa, N.G. (1990) *Colony-Stimulating Factors: Molecular and Cellular Biology*. Marcel Dekker Inc., New York.
20. Gough, N.M. and Williams, R.L. (1989) *Cancer Cells,* **1,** 77.
21. Lyons, R.M. and Moses, H.L. (1990) *Eur. J. Biochem.,* **187,** 467.
22. Barnard, J.A., Lyons, R.M. and Moses, H.L. (1990) *Biochim. Biophys. Acta,* **1032,** 79.
23. Nilsen-Hamilton, M. (1990) *Curr. Top. Dev. Biol.,* **24,** 95.
24. Tracey, K.J., Vlassara, H. and Cerami, A. (1989) in *Peptide Regulatory Factors*. Edward Arnold, London, p. 60.
25. Balkwill, F.R. (1989) in *Peptide Regulatory Factors*. Edward Arnold, London, p. 50.

Further reading

General

Meager, A. (1990) *Cytokines*. Open University Press, Milton Keynes.

Growth factors

Green, A.R. (1989) Peptide regulatory factors: multifunctional mediators of cellular growth and differentiation. in *Peptide Regulatory Factors*. Edward Arnold, London, p. 1.

Ross, R. (1989) Platelet-derived growth factor. in *Peptide Regulatory Factors*. Edward Arnold, London, p. 70.

Waterfield, M.D. (1989) Epidermal growth factor and related molecules. in *Peptide Regulatory Factors*. Edward Arnold, London, p. 81.

Hematopoietic growth factors

Young, N.S. (1990) Hematopoietic growth factors: a summary of their biology in tissue culture, animals, and clinical trials. *Prog. Clin. Biol. Res.* **337,** 539.

Interferons

De Maeyer, E. and De Maeyer-Guignard, J. (1988) *Interferons and Other Regulatory Cytokines*. Wiley, New York.

Interleukins and other lymphokines

Dawson, M.M. (1991) *Lymphokines and Interleukins*. Open University Press, Milton Keynes.

O'Garra, A. (1989) Interleukins and the immune system 1. in *Peptide Regulatory Factors*. Edward Arnold, London, p. 32.

O'Gara, A. (1989) Interleukins and the immune system 2. in *Peptide Regulatory Factors*. Edward Arnold, London, p. 42.

Transforming growth factors

Massagué, J. (1990) The transforming growth factor-β family. *Ann. Rev. Cell Biol.*, **6**, 597.

Salomon, D.S., Kim, N., Saeki, T. and Ciardiello, F. (1990) Transforming growth factor-α: an oncodevelopmental growth factor. *Cancer Cells*, **2**, 389.

Tumor necrosis factors

Larrick, J.W. and Wright, S.C. (1990) Cytotoxic mechanism of tumor necrosis factor-α. *FASEB J.*, **4**, 3215.

2
CYTOKINE GENES AND PROTEINS

2.1 Structure and expression of cytokine genes

The behavior of cytokines in health and disease depends ultimately on the structure and expression of the genes that encode them. As with most other genes of eukaryotes, sequences expressed in the mature mRNA (exons) are usually interspersed with introns, sequences that are transcribed but then spliced out of the RNA during maturation in the nucleus [1] (*Figure 2.1*). Different cytokine genes contain different numbers of exons (*Table 2.1*). The genes for the α and β IFNs are unusual in that they lack introns. The IFNγ gene, however, possesses three introns.

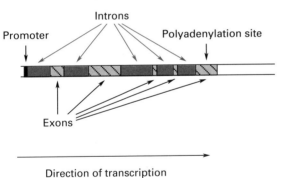

Figure 2.1: Structural features of genes in eukaryotic cells. A typical eukaryotic gene consists of exons (hatched) interspersed with introns (open) that are removed from the primary transcript by splicing. Transcription begins near the 5' end of the gene; a polyadenylate tract is added to the 3' end of the mRNA after transcription.

Upstream of the transcription initiation site, (i.e. at the 5' end of the gene) are sequences that control expression of the gene. These short stretches of nucleotides function as promoters, enhancers and response elements (*Table 2.2*), which bind proteins involved in transcription or its control. Of particular interest are the response elements which enable the gene to be switched on or off in response to regulatory factors (*Table 2.3*). Such mechanisms are essential for genes that regulate rapid physiological responses to changes in the environment.

At the 3' end of the gene, downstream of the protein coding region, lie signals which control transcription termination and the concomitant addition of a poly(A) tail [1]. The polyadenylation signal is the sequence 5' AAUAAA 3'. In some genes there are several polyadenylation sites, allowing the possibility of alternative positions for the poly(A) tail. Such alternative structures may influence the stabilities or translational efficiencies of the mRNAs.

Table 2.1: Exon organization of some cytokine genes

Gene	Number of exons
PDGF (A and B chains)	7
IL-3	5
M-CSF	8
G-CSF	5
TNF (α and β)	4
IL-1 (α and β)	7
IL-2	4
IL-6	5
IFNα (all loci)	1
IFNγ	4

Table 2.2: Control regions that regulate eukaryotic gene transcription

Control region	Definition
Promoter	The site for initiation of transcription
Enhancer	Sequence of DNA that increases the rate of transcription, some distance from the promoter
Response element	Sequence of DNA that mediates changes in the rate of transcription in response to external influences on the cell

Table 2.3: Transcriptional response elements

Name	DNA sequence[a]	Regulatory factors to which it responds
TRE	5' TGAGTCAG 3'	Activators of protein kinase C (acting through products of *FOS* and *JUN* proto-oncogenes)
CREB	5' TGACGTCA 3'	Cyclic AMP-dependent protein kinases
ISRE	5' GGAAAN(N)GAAAC 3'	IFNs (perhaps acting through changes in protein phosphorylation)

[a] The DNA sequences shown are typical 'consensus' sequences. Variations exist, which may or may not affect the strength of the transcriptional activation. N = any nucleotide; (N) = presence or absence of any nucleotide is tolerated.

2.1.1 Gene structure

The primary nucleotide sequences of cytokine genes often suggest relationships between different cytokines. Similarities of exon–intron organization or nucleotide sequence homologies can indicate evolutionary relationships between cytokines and also potential similarities in function. Of course, once the sequences of a gene and its mRNA are known, the primary sequence of the protein it encodes can be predicted. Because of the degeneracy of the genetic code, related proteins often show greater resemblances in amino acid sequence than in nucleotide sequence. Knowing the DNA or mRNA sequence does not always tell us everything about the primary structure of the functional protein product, however, because proteins are often extensively modified after translation by proteolytic cleavage and other changes (see Section 2.2).

The chromosomal locations of cytokine genes (*Table 2.4*) sometimes suggest evolutionary (or even regulatory) relationships [2]. For example, the long arm of chromosome 5 carries the genes for interleukins 3, 4 and 5 as well as GM-CSF and M-CSF. Two cytokine receptor genes (for the M-CSF and PDGF receptors) are also closely linked in this region. The significance of this is not yet clear. In contrast, although EGF and TGFα share structural features (30% similarity at the amino acid level) and a common receptor, the genes for their precursors are quite distinct and located on separate chromosomes. TNFα and β have no more sequence homology than EGF and TGFα, but their genes are closely linked on chromosome 6 (within the MHC gene complex). The genes for the insulin and IGF-II precursors are closely linked on chromosome 11 while that for IGF-I is on chromosome 12, although all three have overlapping abilities to bind to their respective receptors.

Cytokines that consist of more than one sub-unit are not necessarily derived from a single precursor protein (as with insulin), or even from products encoded on the same chromosome. The A and B chains of PDGF show 60% sequence similarity and associate in all possible combinations to give AA, BB and AB dimers, but are coded by genes located on chromosomes 7 and 22, respectively. Similarly with TGFβ, the β1, β2 and β3 forms of the β sub-unit that exist predominantly as homodimers are encoded

Table 2.4: Chromosomal locations of human cytokine genes

Gene	Located on chromosome number
PDGF A chain	7
PDGF B chain	22
EGF	4
Acidic FGF	5
Basic FGF	4
TGFα	2
TGFβ1	19
IGF-I	12
IGF-II	11
NGF	1
IL-3, M-CSF, GM-CSF, IL-4, IL-5	5
G-CSF	17
EPO	7
TNF (α and β)	6
IL-1 (α and β)	2
IL-2	4
IL-6	7
IFNαs (23 loci) and IFNβ	9
IFNγ	12

by genes on different chromosomes, but show extensive homology at the protein and DNA levels. Probably the related genes originated by sequence divergence from a common ancestral gene, and migrated relatively recently to different chromosomal locations. Presumably the often strong conservation of structural features at the protein level reflects the importance of these parts of the molecules for the action and/or regulation of the cytokines.

Many cytokine genes are members of large families that code for a range of proteins. With some (e.g. α IFNs) all members of the family have similar or identical effects. In other cases the functions of family members have diverged extensively, and it is only features like exon–intron organization or the presence of some invariant amino acids that indicate common ancestry. FGFs belong to a gene family that also contains a number of proto-oncogenes [3] (see Chapter 5). The *INT2, HST* and *FGF5* oncogenes (*Table 2.5*) code for mitogenic proteins which contain an FGF-like region. There is also some sequence similarity between FGFs and IL-1. These observations again emphasize the biological relationship between growth factors and proteins able to transform normal cells into tumorigenic cells. The TGFβ family have rather more diverse functions [4] (*Table 2.6*). The bone morphogenetic proteins (BMPs) have up to 40% sequence similarity with TGFβ1 and TGFβ2, and appear to be involved in formation of bone and cartilage. Other members of this family are the inhibins, which regulate interactions between the gonads and pituitary gland (e.g. inhibiting secretion by the pituitary of follicle-stimulating hormone), and the activins, which antagonize these effects. Müllerian inhibitory substance is a testis-specific member of the same family of polypeptides and appears to function during embryonic development of the male reproductive system.

Some cytokine genes can be grouped on the basis of exon–intron structure even though their products perform quite distinct functions. The genes for IL-2, IL-4, IL-5, IFNγ and GM-CSF share discernible features (*Figure 2.2*). In addition these genes are all expressed in activated T lymphocytes, suggesting that aspects of their structure may be important for the common regulation of transcription or RNA processing, rather than for the functions of the proteins. Another pair of cytokines encoded by genes with very similar exon–intron organizations, albeit located on different chromosomes (*Table 2.4*), are IL-6 and G-CSF. If introns help regulate transcription and RNA splicing, then these aspects of gene organization are important, but at present we do not understand why, for example, the M-CSF gene contains eight exons spread over 22 kb of DNA and can produce distinct mRNA species by differential splicing.

2.1.2 Transcription of cytokine genes

Cytokine gene expression is probably regulated at several levels, including modulation of RNA processing, changes in stability of mature mRNAs and effects on the efficiency

Table 2.5: Members of the FGF gene family with mitogenic activity

Gene	Amino acids	Length of region of homology with FGFs (amino acids)
Acidic FGF	155	
Basic FGF	155	
INT2	231	155
HST	206	155
FGF5	267	155

Table 2.6: The TGFβ gene family

Name	Function
TGFβ1, TGFβ2, TGFβ1.2, TGFβ3	Growth regulation; connective tissue synthesis; chemotaxis, etc.
BMP-2A, BMP-2B, BMP-3	Bone morphogenesis
Inhibins A and B	Regulation of the pituitary–gonad axis
Activins A and AB	Antagonism of actions of the inhibins
Müllerian inhibitory substance	Development of male reproductive system

of mRNA translation, but much evidence indicates that transcription is a major control point. Factors that regulate transcription and features that determine its rate are therefore of great interest, especially since producer cells can switch expression of these genes on and off. Cytokine genes are usually transcriptionally inactive in cell nuclei and need a specific stimulus to overcome this repression.

Relatively little is understood about the mechanisms by which inducers cause expression of cytokine genes. Probably they involve many of the signal transduction processes described in Chapter 3. In many cases the ultimate result is probably activation, perhaps by protein phosphorylation, of pre-existing regulatory proteins that bind to response elements close to or within cytokine genes.

2.2 Cytokine production

By their very nature cytokines must only be produced under appropriate circumstances in the body. Thus we find that transcription and translation of cytokine genes and mRNAs are under tight control in normal cells. Cells do not possess stores of cytokine molecules waiting to be secreted in response to a stimulus (as happens with some hormones such as insulin). Nevertheless, many cytokines can be produced rapidly by *de-novo* synthesis.

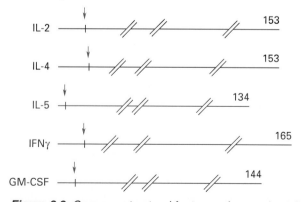

Figure 2.2: Common structural features of several cytokine genes. These five human cytokine genes all have three introns (shown as interruptions in the line) in similar positions. The sizes of individual introns vary and are not shown. Numbers indicate the length of the primary translation product in amino acid residues. Arrows show the cleavage sites for removal of the N-terminal signal sequences initially present in these secreted proteins.

Since all cytokines must exert their effects through receptors on their target cells, they must be efficiently secreted immediately after synthesis. This requires the presence of signal sequences on the primary translation products, allowing them to be sequestered in the lumen of the endoplasmic reticulum (ER) and passed on to the Golgi complex and the secretory pathway [5]. During their passage through these organelles, post-translational modifications can occur, possibly including glycosylation, addition of fatty acid chains and processing of the polypeptide to its final size. Some cytokines (e.g. IL-1) may remain anchored in the plasma membrane of the producer cells and exert their effects by direct cell–cell contact with their target cells [6] (see *Figure 5.2*). This is exceptional, however; most cytokines appear in soluble form in body fluids.

Secreted cytokines are not always active in the form in which they first appear. In some cases precursor molecules are cleaved extracellularly to produce the active protein. Alternatively, a cytokine can be complexed with a carrier or inhibitor protein, and may bind to its receptor only after dissociation of the complex. Clearly, these mechanisms can also potentially regulate cytokine function independently of the rate of synthesis.

2.2.1 Cell specificity

Some cytokines, such as IL-1 and IFNα, are produced by a wide variety of cell types and in response to a range of stimuli. Others are produced by only a single cell type or in response to a very specific stimulus. Several of the interleukins (notably IL-3, IL-4 and IL-5) are largely, if not exclusively, produced by activated T lymphocytes following stimulation by antigens. The basis of such specificity may lie in the structure and regulation of the genes for the individual cytokines. If transcription absolutely requires a *trans*-acting factor that is synthesized only in certain cell types, then gene expression will inevitably be restricted. Similarly, a cell may induce synthesis of a particular cytokine only if it possesses the receptor that allows it to respond to a specific inducer inside or outside the cell. T lymphocytes recognize foreign antigens presented to them by other cells (e.g. macrophages) through the T-cell receptor complex [7]. Thus specific induction of synthesis of some interleukins may be associated with a particular combination of signals generated by this receptor. In contrast, the induction of α and β IFNs by viruses may involve recognition of the early stages of viral replication (e.g. synthesis of double-stranded RNA as an intermediate in RNA virus replication) and this pathway may be common to all cell types that can be infected with such viruses.

Molecular mechanisms by which inducing signals lead to expression of cytokine genes are poorly understood. However, analysis of transcriptional regulation of these genes is progressing rapidly as response elements are identified near to the genes, and the proteins that bind to them are characterized – Chapter 3 discusses this in more detail. For example, recent studies have analyzed in detail the regions of certain IFN genes which permit induction by viruses and double-stranded RNA, and evidence is accumulating that implicates protein phosphorylation in the control of gene expression [8]. Cytoplasmic processes that control cytokine production are also being investigated. At least some cytokine mRNAs may be under a form of translational control and their stability may also be regulated. Several cytokine mRNAs have unusually short half-lives and in some cases sequence elements have been identified that are responsible for this instability [9]. AU-rich sequences in the 3' untranslated regions of the mRNAs for IL-1, IFNβ, IL-6 and GM-CSF appear to destabilize them. Presumably these sequences bind factors that control RNA turnover; such factors (not yet identified) could thus play important roles in controlling the amount of a particular cytokine mRNA and hence the cell's ability to synthesize the corresponding protein.

2.2.2 Secretion of cytokines

By definition, a cytokine should be able to act on target cells distant from the producing cells. This of course requires the cytokine to be secreted by the cells that synthesize it. Generally, all cytokines do appear in secreted form outside their cells of origin, but some can also exist in plasma-membrane-bound forms, from which they are cleaved to release them (*Figure 2.3*, *Table 2.7*). These cell-associated cytokines are significant because there is evidence that the membrane-bound as well as the secreted proteins can stimulate target cells via the usual interactions with appropriate receptors. This requires that the producer and target cells lie in close proximity within a tissue and provides a way of localizing the effect of a cytokine in a truly paracrine or autocrine manner. Such cell–cell interactions probably help regulate hematopoiesis in bone marrow, where stromal cells producing hematopoietic growth factors directly contact the blood cell progenitors that respond to them (see Chapter 5).

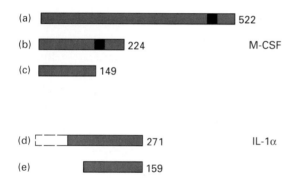

Figure 2.3: Relationships between soluble and membrane-bound forms of cytokines. M-CSF is synthesized as a 522-amino acid (**a**) or a 224-amino acid (**b**) precursor following removal of the N-terminal signal sequence. In transmembrane forms the dark region spans the plasma membrane. Proteolytic cleavage removes the membrane-spanning domain, releasing a 149-amino acid soluble protein (**c**). With IL-1, the 271-amino acid primary translation product (**d**) is processed from the N-terminus and associates with the plasma membrane via a post-translationally attached myristyl group. N-terminal processing releases the 159-amino acid soluble IL-1α (**e**).

Table 2.7: Membrane-bound forms of cytokines

Cytokine	Membrane-bound form (amino acids)	Secreted form (amino acids)
EGF	1217	53
TGFα	160	50
IL-1a	c. 200 (?)	159
TNFα	233	157
M-CSF	224/522[a]	149

[a] Different sizes of the M-CSF precursor may exist; both large and small forms contain a C-terminal membrane-spanning segment.

Secretion of some cytokines via membrane-bound intermediates or other unusual routes may explain why they lack the signal sequences normally required to channel secretory proteins to the exterior. It also introduces another level of complexity into regulation of cytokine production. Specific proteolytic cleavage is necessary to release the soluble cytokine molecules and the proteases could themselves be activated or inhibited by other effector molecules.

2.2.3 Activation of precursors

It is striking that so many of the known cytokines are synthesized as larger precursors which require proteolytic cleavage at specific residues to yield the biologically active molecule (*Table 2.8*). Relatively little is known about the enzymes involved in activation, but probably they are themselves regulated by the producer cells or by more distal influences (e.g. at the site of action of the cytokine).

Table 2.8: Precursor forms of some cytokines

Cytokine	Precursor form (amino acids)	Mature protein (amino acids)
PDGF A chain	211	125
PDGF B chain	226	160
NGF	307	118
IGF-I	130	70
IGF-II	180	67
TGFβ1	391	112

In some cases the mature polypeptide remains non-covalently associated with the cleaved portion of the precursor, and becomes activated only after dissociation [10]; TGFβ provides an example (*Figure 2.4*). This protein is secreted mainly as a complex which cannot bind to receptors and is inactive. Experimentally, activation can occur following brief treatment with acid or alkali, but *in vivo* it is probably achieved by further proteolysis. A third protein associated with the TGFβ precursor complex may be the protease required for activation. Alternatively, activation may take place at sites where TGFβ is required, as in damaged tissues where wound repair is needed. The proteolytic enzymes plasmin and cathepsin D have been shown to activate TGFβ *in vitro*. Clearly, activation by enzymes found only at sites where TGFβ is required would produce the necessary specificity of action of this widely distributed cytokine.

Another way in which the activity of circulating cytokines can be modulated is by covalent or non-covalent association with binding proteins present in serum. Non-covalent association can be tight because of a high affinity of the proteins for each other. The IGFs are largely complexed in this way [11]. IGF-I and IGF-II have combined concentrations in plasma about 1000 times that of insulin, so in order to inhibit a hypoglycemic effect *in vivo* it is important to prevent this large excess of ligand from binding to both insulin and IGF receptors. This is achieved by means of the IGF-binding proteins, which probably also protect the IGFs from degradation, prolonging their half-lives. Other functions are also possible, since IGF-binding proteins in fact constitute a rather heterogeneous group.

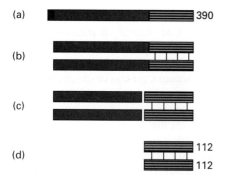

Figure 2.4: *Post-translational activation of TGFβ. TGFβs are synthesized as 390-amino acid precursors (**a**). A short N-terminal signal sequence (dark area) is removed and the proteins dimerize through four disulfide bonds in the C-terminal region (**b**). Proteolytic cleavage generates mature TGFβ (hatched); the large fragment remains non-covalently associated (**c**) and the molecule is inactive. Activation requires dissociation of the 2 x 112-amino acid TGFβ dimer from the complex (**d**). This may be achieved under pH extremes or by further protease activity.*

Soluble forms of some cytokine receptors have been identified [12] and these may also function as binding proteins if they retain their affinity for the ligand. Again, a potential for control can be envisaged if the soluble forms of the receptor compete with the functional cell surface form for limiting amounts of cytokine.

2.2.4 Physiological regulation of cytokine production

Synthesis and secretion of cytokines is usually very closely regulated by environmental factors that influence producer cells, as emphasized elsewhere. Cytokines are produced in response to stimulation of cells by many different ligands, including other cytokines (see *Tables 4.1* and *4.3*). In many cases a single stimulus (e.g. antigen presentation to T lymphocytes) induces a single cell type to synthesize several factors, suggesting that their genes possess common regulatory elements that respond to a single intracellular signal transduction pathway. Much research is currently devoted to understanding these genetic elements and their activation. Interestingly, the signal transduction pathways which regulate cytokine synthesis are often the same as those used by target cells to respond to cytokine stimulation, such as activation of protein kinase C and changes in intracellular calcium levels. Sometimes synthesis of other proteins is needed first and these mediate activation of the cytokine genes. Despite this regulation of their production, some cytokines may possibly be produced constitutively at low levels; for example, some cell types may produce IFNs continuously at a low rate. However, it can be difficult to distinguish a genuine basal level of unstimulated cytokine gene expression from that due to the constant presence of small amounts of an essential inducing agent.

Just as important as the ability to synthesize a cytokine in response to an inducer is the ability of a cell to switch off production when the physiological need has diminished. This is often achieved by sophisticated control mechanisms which not only shut down gene transcription (e.g. by the actions of rapidly turning over repressor proteins), but also cause swift degradation of the mRNA. Some structural features of cytokine mRNAs appear to limit their ability to be translated into protein. For example, as indicated earlier, the mRNAs coding for GM-CSF and IFNβ contain AU-rich regions in their 3′untranslated sequences that cause rapid turnover of these molecules. The net effect of these mechanisms is to allow only transient production of the cytokine in response to an inducer (even when the inducer remains present in the environment of the cell). The limited window of time in which a cytokine can be produced, coupled with the usually short time over which it can act on its target (due, for example, to ligand-induced down-regulation of receptors), means that the physiological effect of a particular inducer *in vivo* will be strictly limited. This is an important aspect of homeostasis (see Chapter 5). There is a further level of complexity in regulation of cytokine production because some of these factors can control their own synthesis in both positive and negative ways. For example, IL-1 induces expression of its own gene [13]; however, stimulation of cells by IL-1 also leads to synthesis of prostaglandin E2, which inhibits IL-1 production.

Equally important is the requirement that cytokines be produced only by the appropriate cell types at the correct locations within the body. This is achieved partly by the nature of the receptors possessed by different cell types, but also by the intrinsic ability or inability of genes to be expressed in different cells. Even within a single organ cells differ in their ability to produce a given cytokine; for example, less than 1% of spleen cells express the IL-4 gene after stimulation with mitogens or antigens [14]. During development and differentiation of multicellular organisms, mechanisms such as DNA methylation lead to permanent inactivation of various genes, ensuring that inappropriate ones cannot be activated. Exactly how this happens, and how specificity is achieved, remain unknown. The differential ability of tissues and cells to produce the various cytokines is believed to play an important role in embryonic histogenesis, morphogenesis and angiogenesis. The specific localization of TGFβ1 to various cellular sites identified by antibody staining techniques [15] in mouse embryos (notably mesenchymal tissues) is an elegant illustration of this point.

2.3 How are cytokines assayed?

Cytokines, like most classical hormones secreted into body fluids, are present in only minute amounts *in vivo*. Their active concentrations are often in the picomolar range since the dissociation constants for high affinity cytokine–receptor binding are usually of the order of 10^{-13} to 10^{-11}M (see *Table 3.1*). This has necessitated development of very sensitive assays to measure their levels and understand their many roles in health and disease.

Since cells are sensitive to such low levels of cytokines, biological assays can often be used. These require target cells that can be cultured *in vitro* and which give an easily measured response. Frequently the response is connected with regulation of cell growth or viability, such as the incorporation of radioactive thymidine into DNA by T lymphocytes in response to IL-2, or lysis of some fibroblast cell lines in response to TNF. IFNs can be assayed using their ability to protect cells against infection by viruses, particularly when the assays are calibrated with standard preparations of known potency. International reference preparations of several IFNs and other cytokines are available, with their biological activities defined in terms of standard units [16].

The difficulty with bioassays is that often more than one cytokine can have the same effect on a target cell, making it impossible to distinguish between different factors that are often present together in a sample. For this reason, as well as the intrinsic

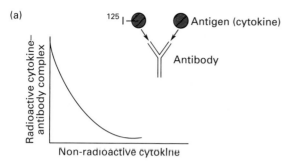

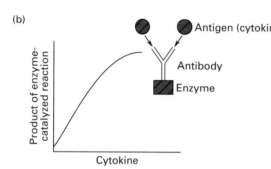

Figure 2.5: Principles of radioimmunoassay and enzyme-linked immunoassay. (**a**) In radioimmunoassay the cytokine to be measured competes with radiolabeled cytokine for a fixed amount of antibody. The concentration of the cytokine is measured by the amount of radiolabel bound in cytokine–antibody complexes. (**b**) Enzyme-linked immunoassay uses antibody coupled to an enzyme such as alkaline phosphatase or peroxidase. The enzyme-catalyzed reaction product is measured after separation of cytokine-bound antibody molecules from excess free antibody.

variability of biological assays, the use of specific antibodies has become the preferred method for assaying cytokines in biological fluids or tissue culture media. Many monoclonal antibodies are now available for radioimmunoassays and enzyme-linked immunoassays (*Figure 2.5*), as well as for immunohistochemical staining of cells and tissue sections and other sensitive immunological techniques [17]. These assays can be highly specific and sensitive, and can often be calibrated accurately using reference standards of known biological potency. Note, however, that assays employing antibodies essentially measure levels of cytokine protein rather than cytokine activity; the two are not necessarily synonymous, particularly where cytokines can be produced in inactive forms or may be complexed with binding proteins that modulate their physiological effects. Such limitations must be kept in mind when examining the evidence for cytokine involvement in disease processes (see Chapter 5).

References

1. Padgett, R.A., Grabowski, P.J., Konarska, M.M., Seiler, S. and Sharp, P.A. (1986) *Ann. Rev. Biochem.*, **55**, 1119.
2. Arai, K., Lee, F., Miyajima, A., Miyatake, S., Arai, N. and Yokota, T. (1990) *Ann. Rev. Biochem.*, **59**, 783.
3. Benharroch, D. and Birnbaum, D. (1990) *Isr. J. Med. Sci.*, **26**, 212.
4. Massagué, J. (1987) *Cell*, **49**, 437.
5. Pfeffer, S.R. and Rothman, J.E. (1987) *Ann. Rev. Biochem.*, **56**, 829.
6. Dainiak, N. (1990) *Prog. Clin. Biol. Res.*, **352**, 49.
7. Clevers, H., Alarcon, B., Wileman, T. and Terhorst, C. (1988) *Ann. Rev. Immunol.*, **6**, 629.
8. Visnavathan, K.V. and Goodbourn, S. (1989) *EMBO J.*, **8**, 1129.
9. Reeves, R. and Magnuson, N.S. (1990) *Prog. Nucl. Acids Res. Mol. Biol.*, **38**, 241.

10. Lyons, R.M. and Moses, H.L. (1990) *Eur. J. Biochem.*, **187**, 467.
11. Ooi, G.T. (1990) *Mol. Cell. Endocrinol.*, **71**, C39.
12. Giri, J.G., Newton, R.C. and Horuk, R. (1990) *J. Biol. Chem.*, **265**, 17416.
13. Warner, S.J.C., Auger, K.R. and Libby, P. (1987) *J. Immunol.*, **139**, 1911.
14. Sideras, P., Funa, K., Zalcberg Quintana, I.Z., Xanthopoulos, K.G., Kisielow, P. and Palacios, R. (1988) *Proc. Natl Acad. Sci. USA*, **85**, 218.
15. Heine, U.I. *et al*. (1987) *J. Cell Biol.*, **105**, 2861.
16. Gearing, A.J.H. and Hennessen, W. (1988) *Developments in Biological Standardization 69 IABS Symposium*. Karger, Basel.
17. Coligan, J.E., Kruisbeek, A.M., Margulies, D.H., Shevach, E.M. and Strober, W. (1991) *Current Protocols in Immunology*. Wiley, New York.

Further reading

Control of cytokine production

Wahl, S.M., McCartney-Francis, N., Allen, J.B., Dougherty, E.B. and Dougherty, S.F. (1990) Macrophage production of TGFβ and regulation by TGFβ. *Ann. NY Acad. Sci.*, **593**, 188.

Cytokine gene transcription

Gasson, J.C., Fraser, J.K. and Nimer, S.D. (1990) Human granulocyte-macrophage colony-stimulating factor (GM-CSF): regulation of expression. *Prog. Clin. Biol. Res.*, **338**, 27.

Roberts, A.B., Kim, S.J., Kondaiah, P. *et al*. (1990) Transcriptional control of expression of the TGFβs. *Ann. NY Acad. Sci.*, **593**, 43.

Tovey, M.G. (1989) Expression of the genes of interferons and other cytokines in normal and diseased tissues of man. *Experientia*, **45**, 526.

Secretion and activation of cytokines

Rutanen, E.M. and Pekonen, F. (1990) Insulin-like growth factors and their binding proteins. *Acta Endocrinol.(Copenh.)*, **123**, 7.

3
HOW DO CYTOKINES WORK?

3.1 Receptors for cytokines

Since all cytokines are proteins or glycoproteins and the plasma membranes of eukaryotic cells are impermeable to virtually all such macromolecules, it follows that cytokines cannot directly enter their target cells. Their effects on cellular functions must therefore be exerted via interactions with structures on the external side of the plasma membrane. These constitute the receptors. The receptors introduce another level at which cytokine action can be regulated, by controlling the amount and/or biochemical activity of the receptors independently of changes in the availability of the cytokines themselves.

The receptors constitute the means by which all the biological effects of the cytokines are brought about. Recent years have seen rapid progress in our understanding of the structure, organization and function of these receptors. Pure cloned cytokine molecules used as labeled ligands bind to cells and help identify the appropriate receptors; this in turn has permitted cloning and characterization of the receptor molecules themselves. Knowing their amino acid sequences has allowed insights into the functions of several cytokine receptors.

In general, a cytokine receptor must consist of at least three parts (*Figure 3.1*). An extracellular domain provides the binding site for the cytokine and creates the specificity for a particular ligand. A transmembrane region spans the phospholipid bilayer of the plasma membrane; and an intracellular domain either has enzymatic activity or binds other molecules, so that a signal is delivered inside the cell in response to the cytokine ligand [1]. Some cytokine or polypeptide hormone receptors (e.g.

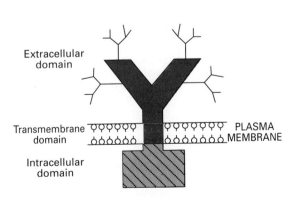

Figure 3.1: Schematic structure of cytokine receptors. Cytokine receptors consist of one or more polypeptide chains spanning the plasma membrane of the cell. The extracellular domain (cytokine binding site) is commonly glycosylated. The hydrophobic transmembrane domain anchors the receptor in the membrane. The intracellular domain is responsible for signal transduction within the cell.

receptors for EGF or PDGF) consist of a single polypeptide chain which can carry out all these functions. In other cases (e.g. the IFNγ receptor) such a single chain can bind the ligand but must interact with a second protein in the plasma membrane or cytoplasm to show biological activity. Finally, the IL-2 and IGF-I receptors are examples where the high affinity ligand binding site is composed of more than one polypeptide chain.

3.1.1 Cytokine binding to cell surface receptors

The specific interaction between a cytokine and its receptor is determined by their three-dimensional structures. Two such large (glyco)proteins can exhibit multiple side-chain interactions, allowing very high affinity and also very specific binding. Binding affinities are usually measured by incubating intact cells or purified receptor with increasing concentrations of the cytokine (often radiolabeled with ^{125}I) and then separating the bound ligand from that remaining free in solution. A Scatchard plot (*Figure 3.2*) gives a straight line if there is a single class of receptor [2]. The dissociation constant (K_d) for the cytokine–receptor complex, and the number of binding sites per cell can be calculated from the plot. *Table 3.1* lists some K_d values. These are typically 10^{-10} to 10^{-12} M, indicating that receptors will be readily occupied by their ligands when cytokines are present at even quite low concentrations in the extracellular environment. The physiological concentrations of many cytokines are often similar to the K_d values of their receptors, as expected if a cell is to respond sensitively to changes in cytokine concentration, since at a ligand concentration equal to the K_d, exactly half the receptors will be occupied.

Scatchard plots of cytokine–receptor interactions often do not produce single straight-line relationships. This can indicate that there is more than one type of binding site for the cytokine. Cells often possess both low affinity (high K_d) and high affinity (low K_d) receptors. Usually only the high affinity receptors modulate cell functions, and usually these are far fewer than the low affinity ones. What physiological significance these large numbers of low affinity, non-functional receptors have is not always obvious; conceivably they could provide a local reservoir of readily available cytokine molecules for rapid occupation of the functional high affinity sites as the free ligand concentration decreased. Another type of relationship can also exist between

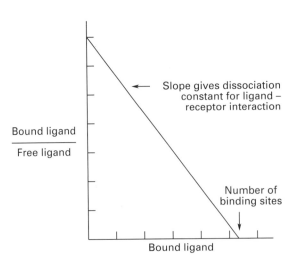

Figure 3.2: Use of the Scatchard plot to analyze cytokine binding to cells. Specific binding is measured over a range of cytokine concentrations and the bound:free ligand ratio plotted against bound ligand. A single class of binding sites gives a straight line from which the dissociation constant (K_d) and number of binding sites can be calculated. If more than one class of binding site is present, more than one straight line may be observed. For further details see reference 2.

Table 3.1: Dissociation constants for cytokine–receptor complexes

Cytokine	Molecular mass of receptor (daltons)	K_d (nM)
EGF/TGFα	170 000	0.1
PDGF		
α receptor	180–185 000	0.01[a]
β receptor	185 000	0.01[a]
FGFs	130–170 000	0.03
IGF-I		
α sub-unit	130 000	1.5
β sub-unit	94 000	
IGF-II	260 000	1
NGF	87–100 000 and 130–140 000 forms	1
IL-1	80 000	0.02–0.25
IL-2		
α sub-unit (Tac)	55 000	0.01 (α + β)
β sub-unit	75 000	10 (β only)
IL-3	140 000	0.0001
IL-4	60 000	0.07
IL-5	92 500	0.01 and 1
IL-6	190 000	0.01–0.1
GM-CSF	180 000	0.001 and 1
G-CSF	150 000	0.0001
M-CSF	150 000	0.0003
EPO	100 000	0.09
TGFβ	300 000	0.01
TNFα/β	75 000 and 95 000	0.0003
IFNα/β	130 000	0.01
IFNγ	90 000	0.01

[a] Depends on the PDGF species bound. The PDGFβ receptor binds tightly only PDGF BB (see Chapter 1).

low and high affinity receptors. A second polypeptide sub-unit may associate with a low affinity receptor to convert it into a fully functional high affinity receptor, as with the IL-2 receptor, where a 55 kd α sub-unit associates with a 75 kd β sub-unit, converting the former from a low (K_d approximately 10^{-8} M) to a high affinity form (K_d approximately 10^{-11}M) [3]. Only the β sub-unit (which has an intermediate affinity on its own) can transmit signals to the T lymphocyte to produce the biological effects typical of IL-2 (*Table 3.2*).

There are also instances of a cytokine being able to bind to more than one type of functional receptor. For example, IGF-I and insulin each bind to the other's receptor (see Chapter 1), although with reduced affinity compared to the homologous ligand (*Table 3.3*). Such cross-reactivity must depend on structural similarities between the two cytokines and between their receptors, but whether this has physiological significance or is just a quirk of evolutionary history is not clear [4]. Recently a second receptor for PDGF has been characterized [5]. The two species of PDGF receptor show some selectivity for the AA and BB forms of this dimeric molecule; the AB heterodimeric form of PDGF, in contrast, is probably a functional ligand for both.

Now that many cytokine receptor genes have been cloned and the amino acid sequences deduced we can recognize superfamilies into which many receptors can be

Table 3.2: Forms of the IL-2 receptor

Sub-unit composition	Molecular mass (daltons)	K_d (nM)
α (Tac protein)	55 000	10
β	75 000	0.8
α+β	55 000 + 75 000	0.02

Table 3.3: Interactions of insulin and the insulin-like growth factors with their receptors

Receptor	K_d (nM)		
	Insulin	IGF-I	IGF-II
Insulin	1.4	100	100
IGF-I	100	1.5	3
IGF-II	—a	10	1

aNo binding of insulin to the IGF-II receptor.

classified. For example, the receptors for PDGF, basic FGF, IL-1, IL-6 and M-CSF have structural similarities and belong to the immunoglobulin superfamily of cell surface proteins [6] (*Figure 3.3*). Similarly, the receptors for IL-2 (β sub-unit), IL-3, IL-4, IL-6, GM-CSF and EPO show some limited homologies and may belong to another superfamily. These weak homologies may indicate evolution from a common ancestral gene, but do not necessarily imply a common mode of action within the cell.

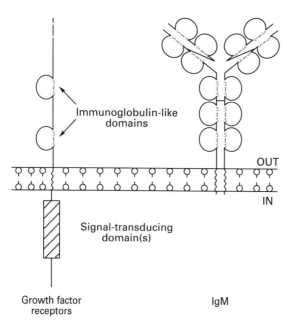

Figure 3.3: The immunoglobulin superfamily of cytokine receptors. The extracellular domains of several cytokine receptors possess features that resemble the structures of immunoglobulin molecules such as IgM, particularly the arrangement of intrachain disulfide bonds. Other cell surface proteins also contain such immunoglobulin-like domains [6].

3.1.2 Receptor internalization and down-regulation

As noted earlier, cytokine activity can be regulated not only by changes in cytokine concentration but also by modulation of its receptor. Both the rate of synthesis and the turnover of the receptor can be regulated. The latter commonly occurs following binding of the ligand, leading to a decrease in receptor numbers on the cell surface. This ligand-induced receptor down-regulation is caused by internalization of ligand–receptor complexes by endocytosis [7] (*Figure 3.4*) and often results in a temporary desensitization of the cell to further stimulation. It has two implications. First, receptor down-regulation may limit the magnitude or duration of the response, preventing a continuing supply of cytokine from over-stimulating the cell. Secondly, internalization of the receptor and/or the cytokine itself may be part of the mechanism by which the cytokine produces its effects. For example, it has been suggested that when IFN enters the cell by receptor-mediated endocytosis, it becomes localized in the nucleus where it may exert direct effects on gene expression [8]. Some support for this controversial idea comes from the finding that human IFNγ synthesized by mouse cells in a non-secretable form can induce various specific cellular responses without interaction with its usual cell surface receptor. This may be exceptional, however; internalization is probably more often the means by which the cell rapidly destroys the ligand by delivery to the lysosomal system.

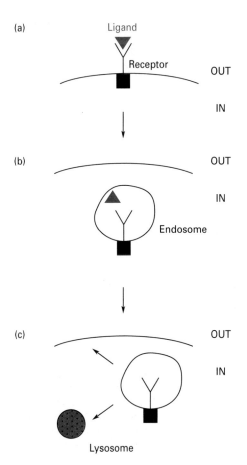

Figure 3.4: Endocytosis of cytokine–receptor complexes. (**a**) A cytokine binds to its receptor. (**b**) The complex is taken into the cell by receptor-mediated endocytosis. The receptor is enclosed within an endosome and the ligand released. (**c**) The cytokine and its receptor are degraded by fusion of the endosome with a lysosome; or the receptor is returned to the cell surface ready to accept another molecule of the ligand.

3.1.3 Receptor recycling and synthesis

The level of cytokine receptors on the cell surface is determined by the balance between the rates of insertion into the plasma membrane and loss due to internalization. Two pathways contribute to assembly of functional receptor molecules at the cell surface – *de-novo* synthesis and recycling of previously internalized receptor (*Figure 3.5*). Receptor synthesis is known to be subject to regulation by a number of influences, but it is less certain whether receptor recycling can also be modulated.

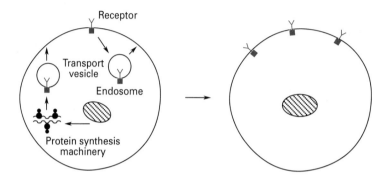

Figure 3.5: *Receptor recycling and synthesis. The level of a receptor on the plasma membrane is determined by the relative rates of internalization (see Figure 3.4) and appearance of recycled or newly synthesized receptor molecules at the cell surface. Increases in receptor number on the cell surface (up-regulation) are usually due to increased* de-novo *synthesis.*

The most dramatic examples of induction of new receptor synthesis are seen in activation of resting lymphocytes by mitogens or antigens. Unstimulated T lymphocytes constitutively express the β but not the α sub-unit of the IL-2 receptor on their surface, giving them an intermediate sensitivity to IL-2 (see *Table 3.2*). Stimulation by IL-1 induces synthesis of IL-2 which in turn induces the α sub-unit of its own receptor [9]. This results in the appearance of the high affinity αβ heterodimer form of the IL-2 receptor, increasing the sensitivity to autocrine stimulation by IL-2 (see Chapter 4). Another example is B-lymphocyte activation, where stimulation of the cells with a ligand such as anti-IgM causes an up to eightfold increase in the number of IL-4 receptors [10]. The contribution of the up- and down-regulation of cytokine receptors to the operation of 'cytokine networks' between different cell types is described in Chapter 4.

3.2 Signal transduction pathways

Many mechanisms are known by which a cytokine or hormone delivered to a cell elicits specific responses. These signal transduction mechanisms form one of the most active areas of modern biochemical research. This aspect of cytokine biology overlaps with more classical areas of biochemical endocrinology; for example, the role of adenylate cyclase and cAMP in hormonal regulation of metabolic pathways has been studied since the 1960s. The concept of 'second messengers' that convey intracellular signals to specific enzymes is therefore an old one, but the field has been widened recently by the discovery of new second messenger molecules. Protein phosphorylation and its regulation is another area where great advances have been made, particularly with the discovery of the tyrosine-specific protein kinases and their role in cellular growth control. These pathways are central to the regulatory actions of cytokines.

3.2.1 Tyrosine kinases and protein phosphorylation

It has long been known that numerous enzymes and metabolic pathways are regulated by reversible phosphorylation of proteins. Cell proliferation appears to be regulated by the tyrosine kinases particularly [11]. These enzymes phosphorylate tyrosines in specific protein substrates, as opposed to many other protein kinases (PKs) which phosphorylate serines or threonines *(Figure 3.6)*. Many different tyrosine kinases have been identified, often because homologs of their genes are actively expressed by a group of tumorigenic viruses, the acutely transforming retroviruses. Several cellular tyrosine kinases turned out to be receptors for growth factors or other cytokines *(Table 3.4)*. In addition, some cytoplasmic tyrosine kinases that do not require ligand binding for activation are nevertheless often associated with the plasma membrane or cytoskeleton of the cell [12]. Examples of the latter are the products of the cellular

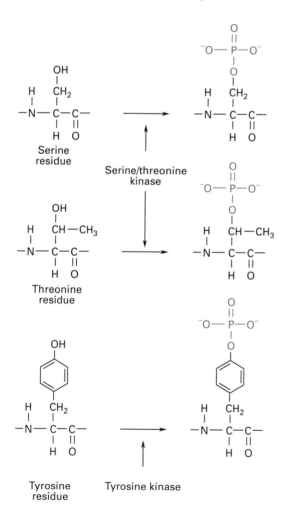

Figure 3.6: Phosphorylation of different amino acids on proteins. The majority of PKs phosphorylate serine or threonine residues; a small group including several cytokine receptors and other plasma-membrane associated proteins, phosphorylate tyrosine residues. In both cases the susceptibility of a residue to phosphorylation depends on the surrounding amino acid sequences.

Table 3.4: Cytokine receptors with tyrosine kinase activity

Insulin-R
IGF-I-R
EGF-R (*ERBB* gene)
NEU (*ERBB2* gene)[a]
PDGF-R
M-CSF (*FMS* gene)
KIT[a]
Atrial natriuretic factor-R

---

[a] Identified as cellular proto-oncogenes.

proto-oncogenes *SRC, FGR, LCK* and *YES*. Cytokine receptors that do not themselves possess kinase activity might regulate activity of associated non-receptor kinases. For example, the IL-3 receptor has no tyrosine kinase activity, but binding of IL-3 to sensitive cells causes increased tyrosine phosphorylation of several proteins [13].

Members of both the receptor and non-receptor tyrosine kinase families have the potential to transform cells to a tumorigenic phenotype. The genes for these proteins are closely related to viral and cellular oncogenes (*Table 3.5*). Uncontrolled cell growth is often associated with mutations, deletions or translocations in these genes that remove ligand binding sites or alter regulation of the kinase activity. In some cases over-expression of the genes may also contribute to tumorigenesis, for example, by rendering cells more sensitive to growth factors.

The protein substrates of the tyrosine kinases have not been well defined, although major targets may be proteins important in cell cycle regulation. Often the enzymes themselves become phosphorylated on one or more tyrosine residues, and cytokine receptors often auto-phosphorylate after binding their ligand. This may be the first event in a cascade of reactions leading to the many biological effects of the cytokine (*Figure 3.7*), as demonstrated for the insulin, IGF-I, PDGF and M-CSF receptors.

Table 3.5: Viral and cellular oncogenes with tyrosine kinase activity

Viral oncogene	Cellular oncogene
v-*erbB*	*ERBB*[a]
v-*fms*	*FMS*
v-*kit*	*KIT*
v-*src*	*SRC*
v-*fgr*	*FGR*
	Other *SRC* family members (*FYN, LYN, HCK, LCK, YES*)
	Non-*SRC* family (*TRK, ROS, FPS/FES, ABL, PIM1, SEA, REL, RET, BCL2*)

[a] Codes for the EGF receptor.

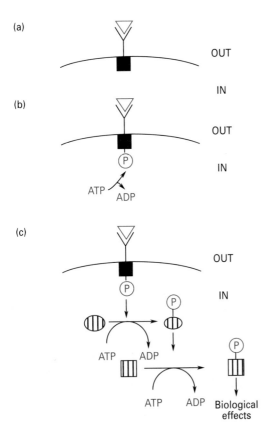

Figure 3.7: Phosphorylation cascades induced by cytokine–receptor binding (**a**) A cytokine receptor with protein tyrosine kinase activity binds its ligand. (**b**) Ligand binding induces auto-phosphorylation which activates the receptor PK, leading to phosphorylation of other substrates on tyrosine residues. (**c**) These substrates may themselves be serine/threonine-specific PKs that phosphorylate a range of targets.

3.2.2 G proteins and receptor-enzyme coupling

Cytokine receptors lacking an intrinsic enzyme activity need a mechanism to couple them to proteins having such functions. Enzyme activation by such a mechanism often results in generation of second messengers that transduce the cytokine–receptor interaction signal to the interior of the cell (see Section 3.2.3). The proteins that couple the receptors to the enzyme systems are GTP-binding proteins (known as G proteins) [14] *(Table 3.6)*.

G proteins are associated with the plasma membrane and consist of α, β and γ sub-units. The α sub-unit binds GDP or GTP and undergoes a cycle of dissociation/reassociation with the β and γ sub-units, driven by binding and subsequent hydrolysis of GTP *(Figure 3.8)*. This in turn regulates the activity of enzymes such as adenylate cyclase, cGMP phosphodiesterase and (possibly) phospholipases in the plasma membrane, with consequences described below.

We know little about possible associations of G proteins with cytokine receptors, but some of the additional proteins reported to co-purify with receptors, or required for full receptor activity, could be components of G-protein complexes. Some growth factor receptors are likely examples. The *RAS* oncogenes [15], whose aberrant expression can contribute to deregulation of cell growth and lead to a tumorigenic phenotype, encode proteins resembling G protein α sub-units, notably in their ability to bind GTP. Over-expression of a normal RAS protein has been shown to couple inositol phosphate production (see Section 3.2.4) to hormonal and growth factor stimulation [16].

Table 3.6: Members of the G protein family

G protein		Molecular mass of sub-units (daltons)	Function
G_s			
	α	45 000	Stimulation of adenylate cyclase
	β	35 000	
	γ	10 000	
G_i			
	α	41 000	Inhibition of adenylate cyclase
	β	35 000	
	γ	10 000	

Other G proteins are believed to exist (e.g. those regulating phospholipase C activity) but are less well characterized.

Conversely, when IFN treatment caused a *RAS*-transformed cell line to revert to normal growth, there was a substantial reduction in RAS protein expression [17].

3.2.3 Adenylate cyclase and cyclic AMP

The best characterized second messenger system in mammalian cells is the well-known adenylate cyclase/cAMP pathway (*Figure 3.9*). cAMP-dependent PK (PK-A) mediates a wide variety of serine/threonine-specific protein phosphorylation events, directly or indirectly. The wide range of PK-A substrates explains why ligands that

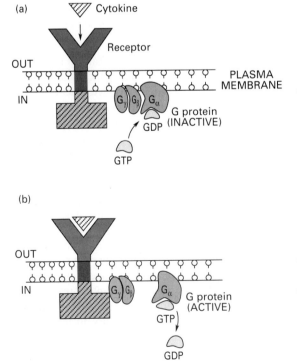

Figure 3.8: Regulation of G-protein activity by guanine nucleotide exchange. (**a**) In the inactive state the three sub-units of a G protein remain associated, and a molecule of GDP is bound to the α sub-unit. (**b**) Following binding of its ligand, the receptor associates with the G protein causing exchange of GTP for GDP on the α sub-unit. The G protein dissociates into a G_γ–G_β dimer and an active G_α–GTP complex. The G_α sub-unit possesses GTPase activity that hydrolyses the bound GTP, allowing reassociation with the other two sub-units and a return to the inactive state.

activate or inhibit adenylate cyclase (via stimulatory or inhibitory G proteins) have so many different effects, and is perhaps a precedent for understanding the multiple effects of other cytokines that use alternative pathways. Surprisingly, few cytokines appear to act directly through cAMP as a second messenger, with the possible exception of IL-1. However, there is evidence for regulation of the adenylate cyclase system by other cytokine-responsive pathways (e.g. by PK-C, which is activated by binding of various cytokines to their receptors). Conversely, PK-A can phosphorylate and regulate proteins involved in other signal transduction mechanisms.

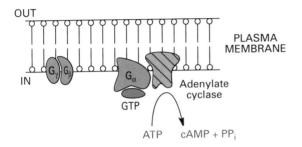

Figure 3.9: Synthesis of cAMP by adenylate cyclase. Adenylate cyclase is regulated by G proteins (see Table 3.6). Following activation of a stimulatory G protein by a hormone–receptor complex (Figure 3.8) G_α–GTP interacts with plasma-membrane-bound adenylate cyclase and stimulates this enzyme to convert ATP into 3'5'cAMP and pyrophosphate.

cGMP has a much more restricted second messenger role than cAMP. It is synthesized by guanylate cyclase and removed by phosphodiesterase-catalyzed hydrolysis to 5'GMP. As yet there are only a few reports suggesting possible involvement of cGMP in cytokine-stimulated signal transduction systems.

3.2.4 Phospholipases

Recently it has become clear that cells possess another group of second messengers, unrelated to the cyclic nucleotides, derived from hydrolysis of membrane phospholipids such as phosphatidylinositol 4,5-bisphosphate (PtdIns(4,5)P_2) and phosphatidylcholine. The enzymes that cleave these substrates are phospholipases (*Figure 3.10*). Best characterized are the phospholipase C enzymes, which hydrolyse the bond between the glycerol and phosphate moieties. Phospholipase C hydrolysis of PtdIns(4,5)P_2 generates two biologically active molecules, inositol 1,4,5-trisphosphate (IP$_3$) and diacylglycerol (DAG) [18]. The latter can also be produced from phosphatidylcholine by the combined actions of phospholipase D and phosphatidate phosphohydrolase (*Figure 3.10*). Both IP$_3$ and DAG act as second messengers and play major roles in regulating intracellular calcium levels and in activating PK-C, respectively, as described below.

There is now good evidence that the activity of phospholipase C enzymes (several of which have been cloned and sequenced) is controlled by a number of cytokines and hormones including IL-1, EGF, and PDGF. Activation may occur via G proteins, as yet unidentified, but at least in the case of EGF and PDGF direct action of the receptors is more likely, since these have the tyrosine kinase activity essential for the activation of phospholipid hydrolysis. At least one form of phospholipase C (γ) undergoes tyrosine phosphorylation in response to EGF or PDGF (but not in response to M-CSF). Conversely, phosphorylation by PKs C or A of serines or threonines in phospholipase Cγ may inhibit its activity.

Another type of phospholipase, phospholipase A_2, cleaves its substrates at a different position (*Figure 3.10*). This can generate arachidonic acid, a precursor of thromboxanes, leukotrienes, prostacyclins and prostaglandins [19] (see Section 3.2.7). IL-1 and TNF stimulate production of these substances, which are believed to be important in the inflammatory and tissue-damaging actions of these cytokines (see Chapters 4 and 5).

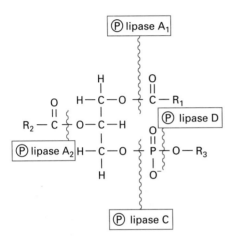

Figure 3.10: Specificities of phospholipases. A typical phospholipid comprises a glycerol residue acylated on carbons 1 and 2 with fatty acids (R_1, R_2) and with a phosphodiester on carbon 3, typically linked to an inositol derivative. Phospholipases A_1, A_2, C and D hydrolyze ester links as shown, generating a variety of potential second messenger molecules.

3.2.5 Inositol phosphates and control of intracellular calcium

Mammalian cells possess many compounds derived from *myo*-inositol, a six-membered ring structure with hydroxyl groups linked to every carbon atom [20] (*Figure 3.11*). The most important of these is IP_3, a second messenger in almost all cell types. IP_3 is produced by hydrolysis of PtdIns(4,5)P_2, catalyzed by members of the phospholipase C family (see Section 3.2.4). It has a transient existence inside the cell, being rapidly dephosphorylated to inositol, but IP_3 plays a very important role in controlling the level of free intracellular calcium ions by causing release of Ca^{2+} from stores within the ER. This increase in Ca^{2+} in turn activates intracellular processes such as glycogenolysis and exocytosis. Many of these effects are mediated by Ca^{2+} binding to calmodulin; this protein can regulate several PKs with a variety of targets. Thus the IP_3–Ca^{2+}–calmodulin pathway controls many cellular processes through phosphorylation-induced changes in the activity of key enzymes (*Figure 3.12*). The other great advantage of this second messenger system, as with the cAMP–PK-A pathway, is that it allows considerable amplification of the initial signal generated by the cytokine binding to its cell surface receptor. Examples of cytokines that activate production of IP_3 are PDGF, EGF and IL-4 (at least in some cell types).

IP_3 can be further metabolized in a different way, by phosphorylation of another hydroxyl group to produce inositol 1,3,4,5-tetrakisphosphate (IP_4) (*Figure 3.13*). This may be just another mechanism for rapid removal of IP_3; alternatively, IP_4 may act as an additional second messenger with a specific role (e.g. potentiating the action of IP_3 in sustaining Ca^{2+} release); its significance is not yet clear. Cells contain even more

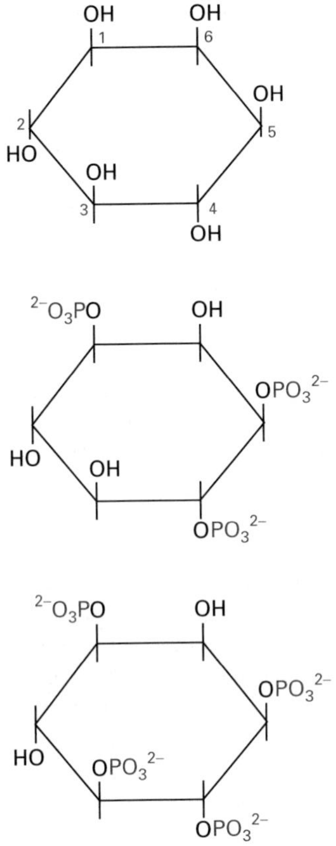

myo-inositol

Ins (1,4,5)P_3
(IP_3)

Ins (1,3,4,5)P_4
(IP_4)

Figure 3.11: Myo-*inositol and its derivatives.* Myo-*inositol can be phosphorylated on each hydroxyl group. The products believed to be of greatest importance in regulating intracellular calcium are IP_3 and IP_4.*

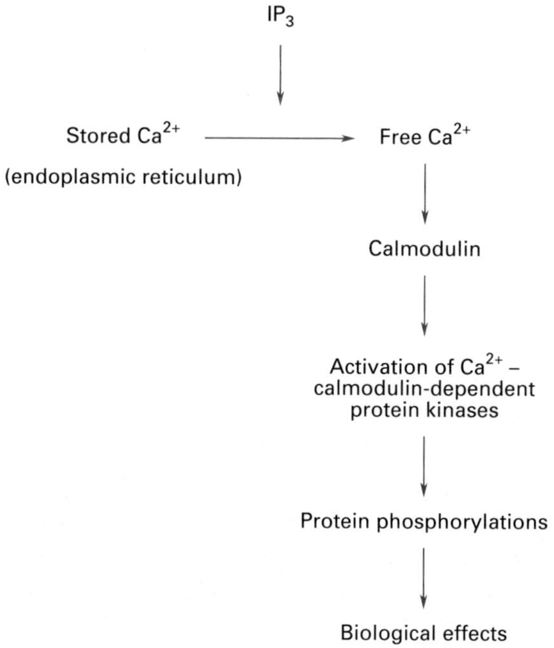

***Figure 3.12:** Mechanism of action of IP_3 as a second messenger.*

phosphorylated derivatives of *myo*-inositol, IP_5 and IP_6, which are synthesized from other inositol phosphates by phosphoinositide kinase and may also function as second messengers in some situations. The β PDGF receptor associates with phosphoinositide kinase and may regulate its activity.

3.2.6 Diacylglycerols and protein kinase C

The other class of second messengers generated by phospholipase-C mediated breakdown of $PtdIns(4,5)P_2$ are the DAGs. These lipids activate the PK-C family of Ca^{2+} and phospholipid-dependent enzymes [21]. PK-Cs comprise at least seven enzymes, probably with subtly different activation requirements and substrate specificities (*Figure 3.14*). These serine/threonine-specific PKs are important regulators of cell proliferation and differentiation. The ability of DAGs to activate PK-C can be mimicked by tumor promoters, of which the phorbol esters are the most widely studied examples. This suggests that PK-C activity must be involved in regulating cell growth, and that perhaps abnormal control of PK-C function can contribute towards the development of tumor cells.

Since hydrolysis of $PtdIns(4,5)P_2$ produces DAGs and IP_3 simultaneously, any cytokine that activates phospholipase C will stimulate both PK-C activity and release of free intracellular Ca^{2+}. The latter will also contribute to activation of PK-C since these enzymes are Ca^{2+}-dependent. Thus PDGF, EGF and IL-4 might be expected to activate PK-C inside target cells. Regulation of PK-C does not necessarily involve phospholipase C; for example, hydrolysis of phosphatidylcholine by phospholipase D generates DAG phosphate (phosphatidic acid) which can yield DAGs after further cleavage by a phosphatase. This has been shown to produce a relatively prolonged elevation of DAGs without any change in IP_3 or free Ca^{2+} concentrations and may be an important part of the signal transduction pathways used by IFNα, IL-1 and IL-3 [22].

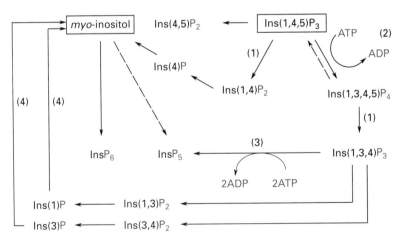

Figure 3.13: *Metabolism of inositol phosphates. Phosphorylated* myo-*inositol derivatives are interconverted by a complex series of reactions. Of particular interest because of their potential regulatory significance are the steps catalyzed by Ins(1,4,5)P$_3$ 5-phosphomonoesterase (1), Ins(1,4,5)P$_3$ 3-kinase (2), phosphoinositide kinase (3) and inositol monophosphate phosphomonoesterase (4). For details see reference 20.*

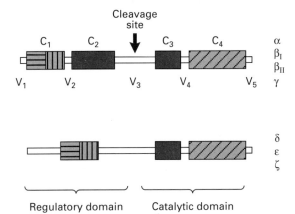

Figure 3.14: *The protein kinase C family. PK-C comprises a family of related enzymes. All possess regulatory and catalytic domains, with a site between them that is susceptible to proteolytic cleavage. C_1–C_4 are regions of conserved amino acids, V_1–V_5 are variable regions. In the α, β_I, β_{II} and γ species of PK-C all these regions are present, whereas in PK-C δ, ε and ζ the C_2 sequence is absent. These subtle differences in amino acid sequence may give rise to different regulatory properties and different substrate specificities for these enzymes.*

3.2.7 Arachidonic acid and prostaglandins

Phospholipids commonly carry arachidonic acid linked to carbon 2 of glycerol. Cleavage of this ester by phospholipase A_2 releases arachidonic acid, a polyunsaturated fatty acid that has important roles in mediating the effects of various cytokines *in vivo* [19]. Arachidonic acid is important because it is a precursor for synthesis of several small lipids known collectively as eicosanoids, which act as inflammatory agents (*Figure 3.15*). The enzyme cyclo-oxygenase converts arachidonic acid into an intermediate that leads to synthesis of prostaglandins. Prostaglandins comprise many species, designated PGA to PGI. They act as local, short-lived signals that influence many activities in the cells that produce them, as well as those nearby [19]. For example, in adipose tissue PGE_1 inhibits hormone-induced increases in cAMP by inhibiting adenylate cyclase. In other cell types, however, adenylate cyclase can be stimulated by prostaglandins. Other effects include control of ion movements across membranes and stimulation of inflammation. Aspirin (acetylsalicylate) and the more potent indomethacin exert their anti-inflammatory effect by inhibiting the cyclo-oxygenase-catalyzed step in the conversion of arachidonate to prostaglandins.

IL-1 and the TNFs activate the conversion of arachidonic acid into eicosanoids. In the latter case increased production of PGE_2 probably follows increased availability of arachidonate due to enhanced phospholipase A_2 activity. The cytotoxic effects of the TNFs can be inhibited by aspirin or indomethacin; however, it is not yet clear whether the cytotoxicity is due to actions of the prostaglandins or other eicosanoids (see Section 3.2.8), or to production of chemically reactive free radicals during the cyclo-oxygenase-mediated reactions. Systemic induction of fever by IL-1, the TNFs and α IFNs can also be attributed to enhanced PGE_2 synthesis, in this case in the hypothalamus. Another molecule, PGI_2 (prostacyclin), is produced by endothelial cells in response to IL-1 or TNFs; it is responsible for the inhibition of platelet aggregation and for the vasodilatory effects of these cytokines that may contribute to physiological shock.

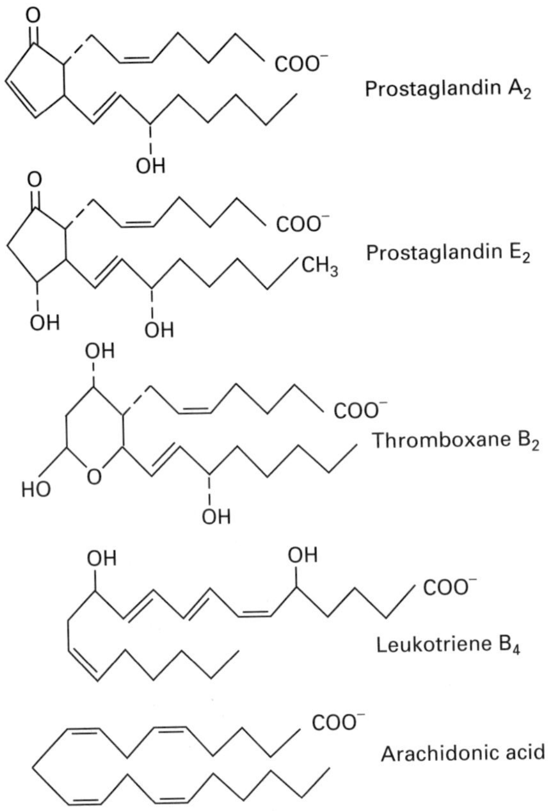

Figure 3.15: Some inflammatory mediators derived from arachidonic acid. Arachidonic acid is a 20-carbon fatty acid with four double bonds which is released from membrane phospholipids by phospholipase A_2. Eicosanoids derived from arachidonic acid include the prostaglandins and thromboxanes (synthesized by the action of cyclo-oxygenase) and the leukotrienes (generated by lipoxygenase).

3.2.8 Production of inflammatory mediators

Like prostaglandins, thromboxanes and leukotrienes are eicosanoids synthesized from arachidonic acid *(Figure 3.15)*. Thromboxane production also requires cyclo-oxygenase (and is therefore inhibited by the same agents that block prostaglandin synthesis), whereas conversion of arachidonate to leukotrienes involves a different enzyme, lipoxygenase. Elevated production of these compounds is associated with inflammatory reactions [23] and is a likely response to any cytokine that activates phospholipase A_2 and arachidonate release.

Phospholipase A_2 has also been implicated in synthesis of another group of inflammatory mediators. Platelet-activating factor (PAF) [24] is a family of compounds structurally related to acetylglycerol-ether-phosphorylcholine, produced by vascular endothelial cells in response to stimulation by IL-1 and the TNFs. The effects of PAF include aggregation of platelets, degranulation of neutrophils and changes in vascular permeability.

3.3 Regulation of gene expression

Ultimately, the behavior of a cell depends on the kinds of proteins that it makes, and advances in molecular biology have permitted identification of new genes whose expression is regulated by cytokines. Two important goals of this work are to discover the mechanisms by which cytokines control expression of the genes, and to understand what the gene products do in the cell.

3.3.1 Transcriptional control mechanisms

It would be beyond the scope of this book to review in detail the mechanisms by which mammalian gene expression is controlled at the DNA level, but briefly, the ability of many genes to be copied into pre-mRNA can be regulated by external factors, and the cytokines clearly constitute one such class of factors. In response to cytokine–receptor interactions at the plasma membrane the following changes are possible within the nucleus.

(a) Transcription of previously silent genes, resulting in appearance of new mRNAs and synthesis of new proteins.
(b) Up-regulation of genes previously expressed at a low constitutive level, leading to elevated levels of the mRNAs and proteins.
(c) Down-regulation of previously highly expressed genes, leading to diminished concentrations of the mRNAs and proteins.
(d) Complete shut-off of expression of certain genes, so that the cell can no longer synthesize the products.

Such transcriptional regulation must involve communication between the plasma membrane and the cell nucleus, via second messengers. Relatively little is known about how this is done, but cytokine-mediated changes in transcription are known to require specific DNA sequences, the response elements, within or close to the individual genes (*Table 2.3*). Response elements most frequently lie just upstream (i.e. on the 5' side) of the promoter, and are targets for *trans*-acting factors, which are proteins that recognize these DNA sequences and bind to them specifically [25]. Such binding can either stimulate or inhibit transcription. Control of these *trans*-acting factors must be the link between membrane and cytoplasmic signalling events and the subsequent changes in gene transcription. Possibly cytokines might stimulate *de-novo* synthesis of some factors, but this would be slow and would require yet another level of transcriptional regulation. It is more likely that the activity of pre-existing DNA-binding proteins is regulated by covalent modification, possibly by phosphorylation using cytokine-responsive PKs such as PK-A or PK-C, Ca^{2+}–calmodulin-regulated PKs or tyrosine kinases. Equally, dephosphorylation by protein phosphatases could regulate the necessary DNA–protein interactions.

If some gene products can influence the expression of other genes, we can imagine a cascade of changes in gene transcription over a period of several hours after the initial cytokine-mediated stimulation (*Figure 3.16*). Thus we can distinguish 'immediate' or 'early' genes, whose transcription is affected within minutes of cytokine treatment, from 'intermediate' or 'late' genes that respond only after several hours (or even days). Operationally, the latter genes do not usually show changes in expression without new protein synthesis, presumably reflecting the production of new *trans*-acting factors. An example is the response of non-proliferating cells such as resting fibroblasts or unstimulated T lymphocytes to growth factors [26] or mitogens [27], respectively. One of the earliest events in the nucleus is the induction of the *FOS* gene, the product of which is itself part of a transcriptional regulatory complex. Combination of the FOS protein with another protein called JUN produces a factor capable of up-regulating transcription of several other genes [28].

As indicated in Chapter 1, transcription of genes encoding cytokines or their receptors can be regulated by the same or other cytokines. This results in subtle interactions and control mechanisms for the effective operation of cytokine networks (see Chapter 4). For example, in T lymphocytes IL-1 may induce transcription of the IL-2 gene by increasing the affinity of transcriptional activators for an enhancer region. For at least one of these activators the stimulation involves PK-C. The transcription

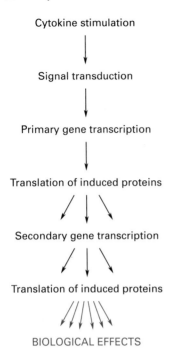

Figure 3.16: Cascades of gene regulation. Many of the ultimate biological effects of cytokines are probably brought about by the products of genes at the ends of regulatory cascades.

factor NF-κB [29] exemplifies another possible way of regulating transcription rapidly. This protein exists in the cytoplasm of cells as a complex with an inhibitory protein, IκB. Only after dissociation of this complex, induced by phosphorylation of IκB, can NF-κB migrate to the cell nucleus and interact with regulatory sequences of genes *(Figure 3.17)*. IκB can be phosphorylated by a number of the PKs (including PK-C) that may transduce cytokine–receptor signals. Activation of NF-κB by this mechanism has been shown to follow treatment of cells with TNFα or IL-1. A similar pathway, but involving a different DNA-binding protein, appears to be involved in induction of gene expression by IFNs [30].

3.3.2 Post-transcriptional regulation

Transcription of a gene is of course necessary for the expression of its protein product. However, many other steps intervene between synthesis of the primary transcript and translation of the mature mRNA into protein. Events such as polyadenylation of the 3' end of the RNA, splicing to remove introns, transport of the mature mRNA to the cytoplasm, and breakdown of the mRNA in the cytoplasm are all potential points where regulation could alter the level of synthesis of the product *(Figure 3.18)*. In addition, the efficiency of translation of the mRNA can change under different circumstances. In principle, cytokines could affect gene expression at any of these stages. In practice there is little evidence for cytokine-mediated regulation of post-transcriptional steps, although this might simply reflect our lack of appropriate experimental approaches. There are good examples of specific translational control in response to other regulatory influences (e.g. iron availability affects translation of ferritin mRNA) [31], and good evidence that the overall rate of translation is controlled by a variable rate of initiation depending on the growth state of the cell. Restoration of essential growth

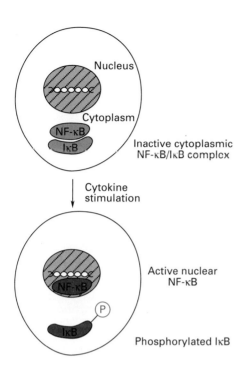

Figure 3.17: Activation of NF-κB in control of gene expression. The transcriptional regulatory factor NF-κB is inactive when complexed with the inhibitory protein IκB. IκB can be phosphorylated by a number of PKs, causing the release of NF-κB.

factors to cells previously deprived increases initiation of protein synthesis. Thus in the future we may anticipate identification of cytokine-sensitive post-transcriptional control mechanisms.

The *MYC* oncogene provides clear evidence for gene regulation at the level of mRNA stability. When cells of the Daudi line, derived from Burkitt's lymphoma, are treated with α or β IFNs, there is a severe inhibition of growth with an increase in the turnover of *MYC* mRNA [32]. The molecular mechanism has not been identified: *MYC* mRNA is unstable even in actively proliferating control cells, and it is not clear whether IFN simply accelerates the normal process of turnover or activates a separate pathway of degradation. Turnover of another oncogene mRNA, *FGR*, may perhaps also be increased in IFN-treated Daudi cells [33]; this mRNA is normally very stable so in this case IFN must specifically activate degradation.

3.4 Control of cell surface proteins

Among the genes regulated by cytokines, some code for proteins expressed on the cell surface. Cell-lineage specific proteins which can be identified through binding of specific antibodies to intact cells are often called differentiation antigens. Another group of such regulated cell surface molecules includes receptors for the same or other cytokines. Some cell surface proteins mediate cell–cell recognition or interaction of the cell with the extracellular matrix. Important examples of such proteins that are subject to control by cytokines are the major histocompatibility complex (MHC) antigens [34] and the cellular 'adhesion molecules'. Cytokine-induced alterations in cell behavior are often mediated through changes in expression of such cell surface proteins.

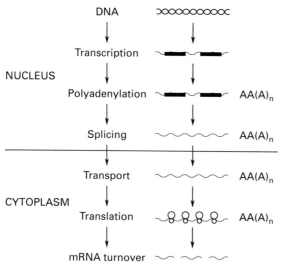

Figure 3.18: Potential control points for post-transcriptional regulation of gene expression.

3.4.1 Differentiation antigens

Exposure of cells to cytokines often induces cell differentiation (see Chapter 4), usually characterized by enhanced or *de-novo* expression of new cell surface antigens. Changes in gene expression must be responsible for such effects, but often the molecular basis has not been investigated. Most studies have been conducted on tumor cell lines which cease proliferation in tissue culture upon treatment with agents such as IFNs or TNF. For example, tumor cells derived from Burkitt's lymphoma, which are normally 'frozen' at a particular stage in the B lymphocyte differentiation pathway, can be induced by α IFNs to express surface antigens characteristic of mature plasma cells [35]. This could explain the therapeutic effect of IFNs in treatment of the rare human 'hairy cell' leukemia, although in this case the effect may be re-direction of differentiation of lympho-myeloid stem cells towards the myelomonocytic rather than the B cell lineage. TNFα can also induce differentiation of human myeloid cell lines and myelogenous leukemias, resulting in conversion to more mature monocytes/macrophages. The differentiation of chronic B-lymphocytic leukemia cells may be responsible for the remissions sometimes seen in this disease following IFN treatment. We do not know how such differentiation is achieved. Normal as well as tumor cells can be induced to differentiate by IFNs. In a clone of cytotoxic T cells IFN was reported to induce rearrangement of the T-cell receptor α chain gene [36], although the mechanism remains unknown.

3.4.2 Major histocompatibility complex antigens

Genes coding for the MHC antigens (*Figure 3.19*) comprise an important group whose expression is regulated in appropriate cell types. These proteins present antigenic peptides ('epitopes') to T lymphocytes for recognition by the T-cell receptor complex. It is through the MHC antigens that the immune system distinguishes between 'self' and 'non-self' antigens. Macrophages, which process foreign proteins and present epitopes for immune recognition, depend on MHC gene expression; accordingly, it is regulated in these cells by several cytokines including IFNγ and TNFα. In fact these two

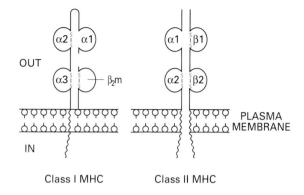

Figure 3.19: *Structure of major histocompatibility antigens. Class I MHC antigens consist of a 45 kd heavy chain anchored in the membrane, with three extracellular domains ($\alpha 1$, $\alpha 2$ and $\alpha 3$), a transmembrane region and a cytoplasmic tail; associated with this is another chain the β_2 microglobulin molecule. Class II MHC antigens consist of non-covalently linked α and β chains that both span the membrane. The positions of intrachain disulfide bridges (–S–S–) are indicated. Carbohydrate chains are also found linked to the extracellular domains.*

factors synergize in inducing class II MHC expression in macrophages. Curiously, in resting B lymphocytes class II MHC expression is inducible by IL-4 but not IFNγ.

IFNs induce genes of the MHC family by transcriptional activation. There are differences between the different IFN species in this ability. IFNγ is much more active at inducing class II MHC genes than are IFNα or β – indeed the latter can sometimes antagonize induction by IFNγ. In contrast, expression of class I MHC genes is stimulated by all IFNs. DNA sequences necessary for transcriptional control by IFNs have been identified upstream of the MHC genes, but until the differences in signal transduction pathways and nuclear *trans*-acting factors regulated by the IFNα/β and IFNγ receptors have been elucidated, the mechanisms underlying these differences in gene responsiveness will not be fully understood.

References

1. Cosman, D., Lyman, S.D., Idzerda, R.L., Beckmann, M.P., Park, L.S., Goodwin, R.G. and March, C.J. (1990) *Trends Biochem. Sci.*, **15**, 265.
2. Dahlquist, F.W. (1978) *Methods Enzymol.*, **48**, 270.
3. Greene, W.C., Wano, Y. and Dukovich, M. (1988) in *Recent Progress in Hormone Research*, Vol. 44 (J.H. Clark, ed.). Academic Press, New York, p. 141.
4. Czech, M.P., Lewis, R.E. and Corvera, S. (1989) *Ciba Found. Symp.*, **145**, 27.
5. Matsui, T., Heidaran, M., Miki, T. *et al.* (1989) *Science*, **243**, 800.
6. Williams, A.F. and Barclay, A.N. (1988) *Ann. Rev. Immunol.*, **6**, 381.
7. Goldstein, J.L., Brown, M.S., Anderson, R.G.W., Russell, D.W. and Schneider, W.J. (1985) *Ann. Rev. Cell Biol.*, **1**, 1.
8. Kushnaryov, V.M., MacDonald, H.S., Sedmak, J.J. and Grossberg, S.E. (1985) *Proc. Natl Acad. Sci. USA*, **82**, 3281.
9. Smith, K.A. (1988) *Interleukin-2*. Academic Press, San Diego.
10. Ohara, J. and Paul, W.E. (1988) *Proc. Natl Acad. Sci. USA*, **85**, 8221.
11. Ullrich, A. and Schlessinger, J. (1990) *Cell*, **61**, 203.
12. Eiseman, E. and Bolen, J.B. (1990) *Cancer Cells*, **2**, 303.
13. Linnekin, D. and Farrer, W.L. (1990) *Biochem. J.*, **271**, 317.

14. Gilman, A.G. (1987) *Ann. Rev. Biochem.*, **56**, 615.
15. Downward, J. (1990) *Trends Biochem. Sci.*, **15**, 469.
16. Wakelam, M.J.O., Davies, S.A., Houslay, M.D., McKay, I., Marshall, C.J. and Hall, A. (1986) *Nature*, **323**, 173.
17. Samid, D., Chang, E.H. and Friedman, R.M. (1984) *Biochem. Biophys. Res. Commun.*, **119**, 21.
18. Wahl, M.I., Nishibe, S. and Carpenter, G. (1989) *Cancer Cells*, **1**, 101.
19. Lewis, R.A. (1990) *Adv. Prost. Thromb. Leuko. Res.*, **20**, 170.
20. Downes, C.P. and Macphee, C.H. (1990) *Eur. J. Biochem.*, **193**, 1.
21. Houslay, M.D. (1991) *Eur. J. Biochem.*, **195**, 9.
22. Reich, N.C. and Pfeffer, L.M. (1990) *Proc. Natl Acad. Sci. USA*, **87**, 8761.
23. Smith, W.L. and Borgeat, P. (1985) in *Biochemistry of Lipids and Membranes* (D.E. Vance and J.E. Vance, eds). Benjamin-Cummings, Menlo Park, p. 325.
24. PaubertBraquet, M., Koltz, P., Guilbaud, J., Hosford, D. and Braquet, P. (1990) *Adv. Exp. Med. Biol.*, **264**, 275.
25. Jones, N.C., Rigby, P.W.J. and Ziff, E.B. (1988) *Genes Dev.*, **2**, 267.
26. O'Keefe, E.J. and Pledger, W.J. (1983) *Mol. Cell. Endocrinol.*, **31**, 167.
27. Crabtree, G.R. (1989) *Science*, **243**, 355.
28. Kouzarides, T. and Ziff, E. (1989) *Cancer Cells*, **1**, 71.
29. Shirakawa, F. and Mizel, S.B. (1989) *Mol. Cell. Biol.*, **9**, 2424.
30. Kessler, D.S., Veals, S.A., Fu, X.-Y. and Levy, D.E. (1990) *Genes Dev.*, **4**, 1753.
31. Aziz, N. and Munro, H.N. (1987) *Proc. Natl Acad. Sci. USA*, **84**, 8478.
32. Dani, C., Mechti, N., Piechaczyk, M., Lebleu, B., Jeanteur, P. and Blanchard, J.M. (1985) *Proc. Natl Acad. Sci. USA*, **82**, 4896.
33. Sharp, N.A., Luscombe, M.J. and Clemens, M.J. (1989) *Oncogene*, **4**, 1043.
34. Widera, G. (1986) *Science*, **233**, 437.
35. Exley, R., Nathan, P., Walker, L., Gordon, J. and Clemens, M.J. (1987) *Int. J. Cancer*, **40**, 53.
36. Chen, L.K., Mathieu-Mahul, D., Back, F.H., Dausset, J., Bensussan, A. and Sasportes, M. (1986) *Proc. Natl Acad. Sci. USA*, **83**, 4887.

Further reading

Adenylate cyclase and cyclic AMP

Mizel, S.B. (1990) How does interleukin 1 activate cells? Cyclic AMP and interleukin 1 signal transduction. *Immunol. Today*, **11**, 390.

Phospholipases, inositol phosphates and calcium

Whitman, M. and Cantly, L. (1989) Phosphoinositide metabolism and the control of cell proliferation. *Biochim. Biophys. Acta*, **948**, 327.

Protein kinase C

Berry, N. and Nishizuka, Y. (1990) Protein kinase C and T cell activation. *Eur. J. Biochem.*, **189**, 205.

Receptor structure and function

Park, L.S. and Gillis, S. (1990) Characterization of hematopoietic growth factor receptors. *Prog. Clin. Biol. Res.*, **352**, 189.

Signal transduction – general aspects

Cantley, L.C., Auger, K.R., Carpenter, C., Duckworth, B., Graziani, A., Kapeller, R. and Soltoff, S. (1991) Oncogenes and signal transduction. *Cell*, **64**, 281.

Vanderhoek, J.Y. (1990) *Biology of Cellular Transducing Signals*. Plenum, New York.

Transcriptional control

Bohmann, D. (1990) Transcription factor phosphorylation: a link between signal transduction and the regulation of gene expression. *Cancer Cells*, **2**, 337.

Tyrosine kinases

Hunter, T. (1991) Cooperation between oncogenes. *Cell*, **64**, 249.

4
BIOLOGICAL ROLES OF CYTOKINES

The cytokines have wide-ranging effects in multicellular organisms since, together with the classical hormones, they co-ordinate the activities of the different tissues and cell types and contribute to homeostasis. How these effects are achieved is rarely understood in detail; for many cytokines such understanding is complicated by their pleiotropic actions on cells and varying spectrum of effects on different cell types. Furthermore, several of the cytokines influence the production and action of other cytokines so that, *in vivo*, it is often hard to tell which is responsible for a given physiological effect. This chapter describes the major biological actions of cytokines, acting individually or in concert, and analyzes current knowledge of the mechanisms.

4.1 Cytokine networks

Cytokines interact with each other in a variety of networks *in vivo*. Several are induced by common stimuli, either in a single cell type or in different cell types *(Table 4.1)*. Often different cell types can produce the same cytokine. Cytokines can stimulate or inhibit synthesis and secretion of other cytokines, either directly or by causing cells to respond differently to other inducing agents. Several cytokines also exert similar or overlapping biological effects *(Table 4.2)*, due to use of common receptors on the target cells (e.g. EGF and TGFα) or to similarities in the intracellular pathways activated by different receptors. Some activities, however, are unique to individual cytokines. The spectrum of effects brought about by a given cytokine can depend both on the nature and on the functional state of the target cell. The latter can of course be modulated by other influences (including other cytokines) which may regulate receptor activity and/or post-receptor pathways inside the cell. At the extreme, a cytokine may be growth inhibitory (or even cytotoxic) for one cell type but mitogenic for another. For example, TGFβ is a growth inhibitor for epithelial, endothelial and hematopoietic cells but a mitogen for some mesenchymal cell types [1]. IL-1 and TNFα have similarly divergent effects in other systems.

A widespread feature of cytokine networks is the ability of cytokines to induce the synthesis of others *(Table 4.3)*. Many additional cases are coming to light as research continues. IL-1, in particular, is a potent inducer of other cytokines, as well as being induced by several agents itself. Some pairs of cytokines, for example, IFNα or β and IL-1; IL-6 and IL-1; GM-CSF and TNFα, can induce each other. Sometimes the full response depends on synergy between the cytokine and an additional activation signal, for example, antigens or mitogens. Such effects may amplify the effect of an initial stimulus on the rate or extent of production of a cytokine *in vivo*. It is not yet clear

Table 4.1: Induction of cytokines by common stimuli

Stimulus	Cytokines induced	Producer cells
Processed antigens/ class II MHC	IL-1 to IL-6 GM-CSF TNFβ	T lymphocytes T lymphocytes T lymphocytes
Specific antigens	IL-1 and IL-6 IFNγ	B lymphocytes B lymphocytes
Viruses and double-stranded RNA	IFNαs IFNβ IL-6 TNFβ	Fibroblasts Fibroblasts Fibroblasts T lymphocytes
Bacterial LPS	IL-1, IL-8 TNFα M-CSF G-CSF GM-CSF IL-7	Macrophages Macrophages Macrophages Macrophages Macrophages Stromal cells

Table 4.2: Overlapping biological effects of cytokines

Biological effects	Inducing cytokines	Cell types
Mitogenesis	EGF, TGFα PDGF FGFs IL-6 IGF-I	Epithelial Mesenchymal Many Plasmacytomas Many
Growth inhibition	TGFβ IFNα/β/γ	Many Many
Induction of MHC antigens	IFNα/β/γ TNFα/β IL-4	Many Many B lymphocytes
Cell differentiation	CSFs EPO IFNα/β TNFβ IL-6 IGF-I/II	Hematopoietic cells Erythroid cells B lymphocytes Monocytic lines Plasma cells Musculoskeletal precursors
Induction of inflammatory mediators	TNFα IL-1 M-CSF GM-CSF TGFβ PDGF EGF	Many Mesenchymal Macrophages Neutrophils Connective tissue Mesenchymal Epithelial

Table 4.3: Induction of cytokines by other cytokines[a]

Cytokines induced	Inducers
IL-1	IL-1; IL-3; IL-4; IL-5; IFNα/β IFNγ; TNFs; GM-CSF; G-CSF; TGFβ
IFNαs	IFNγ; IL-1; M-CSF
TNFs	IL-2; IFNγ; GM-CSF
GM-CSF	IL-1; TNFs; IFNγ; M-CSF

[a] In the examples given here induction of one cytokine by another can either be a direct effect or may require the presence of additional activating signals such as antigens or mitogens.

whether the opposite phenomenon, inhibition of cytokine production by other cytokines, is equally widespread, although one might see the need for such mechanisms to limit cytokine release *in vivo*.

If one cytokine can induce another, then a given effect may be caused at the molecular level by the secondarily induced cytokine rather than the primary agent *(Figure 4.1)*. For example, some of the anti-viral effects of IFNγ may really be mediated by IFNα induced in cells exposed to IFNγ [2]. In cultured cells it may not be hard to determine whether an effect is due to the primary cytokine or to secondary factors, for example, by seeing whether changes in gene expression require new protein synthesis. Neutralizing antibodies against induced cytokines can be used to determine whether they block a given effect. In whole organisms, however, such investigations are rarely

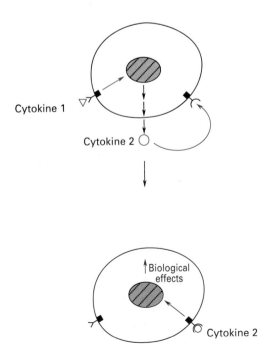

Figure 4.1: Biological effects due to secondary induction of cytokines. Here cytokine 1 induces the synthesis and secretion of cytokine 2. Receptors for cytokine 2 already exist on the cell surface, or are also induced by cytokine 1. Interaction of cytokine 2 with its receptors brings about the biological effects. The effects of cytokine 1 require new protein synthesis, are mimicked by treatment with cytokine 2 and may be inhibited by antibodies that neutralize cytokine 2 activity.

possible because of the difficulty of completely blocking production or activity of an induced protein.

Sometimes a cytokine may not induce the synthesis of another cytokine on its own but can 'prime' cells to become responsive to a co-inducer. For example, IFNγ can prime monocytes and macrophages to produce TNFα or β when stimulated with bacterial LPS or IL-2. IFNγ probably primes IL-1 production under similar circumstances. Production of GM-CSF by T lymphocytes is enhanced by IFNγ, but only in the presence of an activating mitogen [3].

Table 4.4: Transmodulation of cytokine receptors

Receptor	Up-regulated by	Down-regulated by
TNF-R	IFNα/β/γ	TNFα; IL-1
IL-2-R	IFNγ; TNFα; IL-1; IL-2, IL-5; IL-6	
IL-3-R		IL-3
GM-CSF-R		IL-3; GM-CSF; M-CSF
M-CSF-R		IL-3; GM-CSF; M-CSF; G-CSF
G-CSF-R		IL-3; GM-CSF; M-CSF; G-CSF

We do not know the mechanisms underlying all these interesting and important effects, but they probably involve either transmodulation of the levels of receptors for other cytokines *(Table 4.4)* or changes in gene expression which increase the response of the cell when a second cytokine binds to its receptor. The priming effect of IFNγ for TNFα/β production, described above, is accompanied by induction of TNF receptors, ensuring that when the TNFs are produced they can act on their target cells rapidly and with greater effect. IFNγ also induces expression of the IL-2 receptor in monocytes, as do TNFα and IL-1. The mitogenic or growth inhibitory effects of some cytokines may also be due to induction or down-regulation of receptors for other growth promoters. For example, TNFα increases, and IFNα decreases, the number of receptors for EGF on fibroblasts [4]. Such regulation by cytokine networks is superimposed on the down-regulation of receptors induced by binding and internalization of the normal ligand, a widespread phenomenon that transiently desensitizes a cell to further stimulation by that ligand until receptor numbers on the cell surface have recovered (see Chapter 3).

4.1.1 Autocrine and paracrine effects of cytokines

In addition to the functional networks of cytokines that ensure 'cross-talk' between different ligand–receptor systems, there are probably physical networks that allow interaction of cytokines produced by one cell with receptors on adjacent cells. When these nearby cells are of the same type the result will be autocrine regulation by the cytokine; when the target cells are of a different phenotype paracrine regulation can occur (see Chapter 1). These physical contacts allow cell stimulation by high local concentrations of cytokines, while more distant tissues and cell types are only exposed to much lower levels of the factors, giving a graded effect. Indeed, stimulation of target cells can occur with cytokines that are still anchored on the surfaces of producer cells, so that cell to cell signalling requires direct physical contact between the two cell types (see Chapter 2). This may be particularly important in bone marrow where stromal cells can produce CSFs that regulate the activity of neighboring hematopoietic precursor cells [5]. Similarly, proliferation and differentiation of B lymphocytes in active lymphoid follicles will be modulated by local production of interleukins by surrounding T lymphocytes and other cells.

Such autocrine and paracrine effects may be very important for the ability of tumors to grow, invade adjacent normal tissues and metastasize to other sites in the body. They may be responsible for the pathological features of Kaposi's sarcoma (a tumor associated with AIDS) [6]. Factors released by HIV-1-infected T cells or monocytes may stimulate the endothelium-derived spindle cells that are precursors of the tumor cells. The spindle cells in turn produce acidic and basic FGF, IL-1, PDGF, TGFβ and GM-CSF. These cytokines may act not only to stimulate proliferation of the spindle cells themselves but also to promote local angiogenesis, growth of adjacent normal fibroblasts and infiltration by inflammatory cells *(Figure 4.2)*. Such a system may exemplify a common type of mechanism by which a tumor is able to establish a growth advantage in its host.

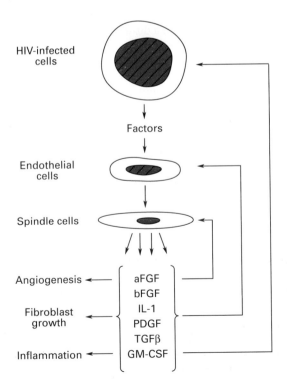

Figure 4.2: Model for contribution of cytokines to development of Kaposi's sarcoma. HIV-infected T lymphocytes and/or macrophages produce cytokines and growth factors that stimulate endothelial cells to change their morphology and phenotype, giving rise to spindle cells typical of Kaposi's sarcoma. This conversion may also involve a genetic change. The spindle cells produce cytokines that exert regulatory influences on the HIV-infected cells and the surrounding normal endothelial cells and fibroblasts. They have angiogenic and inflammatory effects on the adjacent tissues, giving rise to the lesions typical of Kaposi's sarcoma. Adapted from reference 6.

4.2 Regulation of cell growth and differentiation

Many of the effects of the cytokines are of course concerned with controlling cell growth and differentiation. Normal cells will generally proliferate only in response to specific signals from outside factors. Historically, serum was identified as an important source of such factors, and is still widely used experimentally for growing cells in tissue culture. *In vivo*, however, most other body fluids probably also contain growth factors and other cytokines, whose nature and relative amounts may well be very different from those in the bloodstream. In addition to the positively acting growth factors we now recognize several inhibitors among the cytokines (see Chapter 1). Some inhibitors are cytotoxic or simply cytostatic, but others play more dominant roles in promoting terminal differentiation (almost always associated with inability of cells to proliferate further).

The mechanisms by which the factors produce their positive and negative effects on cell proliferation have been intensively studied in recent years. This research effort has depended on cloning, characterization and expression of genes for cytokines and the proteins they regulate, on characterization of cytokines using monoclonal antibodies, and on the realization that many human proliferative diseases involve dysregulation of pathways modulated by growth factors and cytokines. This section provides a summary of the molecular mechanisms by which these agents control cellular growth and differentiation.

4.2.1 Stimulation of cell proliferation by growth factors

Regulation of DNA replication and cell division is best understood in terms of the cell cycle [7] *(Figure 4.3)*. Most cells in most adult tissues are in a non-proliferating 'resting' or quiescent state, often called the G_0 phase of the cell cycle (G is the gap between the observable events of cell division). Proliferating cells move through the G_1, S, G_2 and M phases. Some authors have argued that G_0 is not a distinct state but merely an infinitely long form of G_1, but for non-proliferating cells the switch back into prolif-

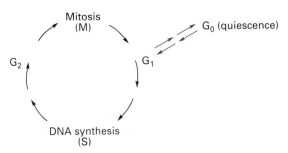

Figure 4.3: *The cell cycle and its regulation. Proliferating cells cycle through four phases: G_1, between the last cell division and a new round of DNA synthesis; S, the synthetic phase in which all the cell's DNA is replicated once in the usual semi-conservative manner; G_2, between DNA synthesis and mitosis; M, mitosis when the chromosomes segregate into the two daughter nuclei and the cell divides by cytokinesis. Quiescent cells are believed to exist in a G_0 phase, from which they must be stimulated to enter the cycle. For cells to become committed to DNA synthesis and to mitosis, a number of events (symbolized by arrows) are required, involving regulatory factors such as the PK $p34^{cdc2}$.*

eration involves specific events in G_0/G_1 that commit the cell to DNA synthesis (often referred to as 'start', and then to allow it to progress to the initiation of S phase. In a cell population that is already proliferating, there can be regulatory points both in passage through G_1 and at the exit from G_2/M. Some of these regulatory steps probably involve protein phosphorylation. Particular interest now centers on the p34^{cdc2} protein. Genetic studies show that this protein is involved in cell cycle control both in G_1 and at mitosis; it has two alternative forms that promote entry into S phase and passage through mitosis, respectively. p34^{cdc2} is a PK that is itself regulated by phosphorylation [8]; its different roles may be performed by differently phosphorylated forms or by different complexes with other factors.

Several growth factors control the switch from quiescence to proliferation by acting as 'competence' and/or 'progression' factors in G_0/G_1. Both types of stimulation are believed to be necessary for a cell to enter S phase. In some cases a single growth factor (e.g. PDGF) may act in both capacities and allow proliferation to commence. More often a full proliferative response requires a combination of growth factors. For example, insulin or IGF-I have been described as progression factors, necessary but not sufficient for mitogenesis. However, these roles are not necessarily fixed, as the effects of a given factor will also depend on the cell type studied.

The distinction between competence and progression must lie in the nature of the genes that are regulated by different growth factors. Certain genes are induced rapidly after exposure of quiescent cells to mitogenic factors, and some of their products must be important in growth control, because some acutely transforming retroviruses owe their tumorigenic phenotype to the presence of mutated forms of these genes. These are the oncogenes [9]. The normal cellular equivalents of viral oncogenes (strictly proto-oncogenes, but often called simply oncogenes) are also often over-expressed in tumor cells. Expression of many proto-oncogenes is regulated by growth factors during the switch from cellular quiescence to proliferation [10]. The *FOS* and *MYC* proto-oncogenes are induced within minutes of treating target cells with PDGF or other mitogenic cytokines. This effect is often transient, suggesting that the products of these 'early' genes rapidly activate expression of other genes that themselves control proliferation. Thus the FOS protein is part of a transcriptional regulatory complex able to induce other genes [11], and the MYC protein may have a related role [12].

A cell must need many gene products to enter the S phase and traverse the cell cycle. Enzymes for biosynthesis of DNA precursors and for DNA replication (e.g. ribonucleotide reductase, thymidine kinase and DNA polymerases) may have to be induced. Transport systems for uptake of nutrients (e.g. the plasma membrane glucose transporter), and all the protein biosynthetic machinery (e.g. ribosomes) must be present in adequate amounts to support cell growth. However, these materials alone cannot switch a cell from a non-proliferating to an actively dividing state; probably they are themselves regulated by the activity of the more rapidly expressed growth factor-sensitive proto-oncogenes. It is significant that the products of most known proto-oncogenes either are themselves involved in signal transduction pathways normally regulated by cytokines (see Chapter 3) or are rapidly induced by such pathways *(Table 4.5)*. All are likely to exert controlling influences at early stages of the response to mitogens *(Figure 4.4)*.

Recently it has become clear that while the proto-oncogene class of regulatory genes exert a positive effect on cell proliferation, a second group, the anti-oncogenes or tumor suppressor (TS) genes, act to prevent unregulated cell proliferation [13]. Some hereditary tumors such as retinoblastoma result from loss or inactivation of both alleles of TS genes. Some mitogenic growth factors may act by inhibiting TS gene

Table 4.5: *Involvement of proto-oncogene products in signal transduction pathways*

Proto-oncogene	Function in signal transduction
ERB, ERBB2, FMS, KIT	Receptors for growth factors or growth factor-like ligands
SRC, FGR, FYN LCK, HCK, LYN YES, CRK	Plasma-membrane associated proteins that may be linked to receptors for growth factors, and other mitogenic or differentiation-inducing ligands. Members of this group all have protein or tyrosine kinase activity
TRK, ROS, FPS/FES REL, RET, SEA	As for the previous group, but without tyrosine kinase activity
RAS family	G-protein-like functions in regulation of second messenger synthesis (?)
MOS, MIL/RAF	Cytoplasmic PK activity (serine-specific)
FOS, MYC family MYB, JUN	Nuclear-DNA-binding proteins with possible transcriptional regulatory activity

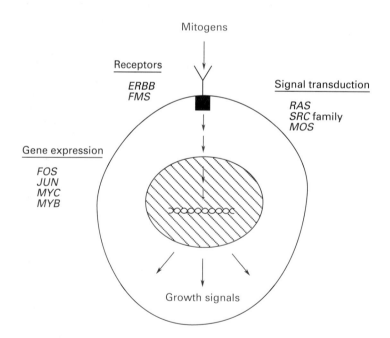

Figure 4.4: *Roles of proto-oncogenes in the response to mitogens. Many cellular proto-oncogenes code for cytokine receptors, enzymes involved in signal transduction (PKs or G proteins), or nuclear regulators of gene expression. Mutations or over-expression can generate growth signals independently of stimulation by mitogens.*

expression or by inactivating TS gene products, perhaps by reversible changes in protein phosphorylation state. Recently the *RB1* (retinoblastoma) gene product has been shown to inhibit expression of the *FOS* proto-oncogene in a manner that depends on the state of phosphorylation of the RB protein [14]. The latter in turn varies through the cell cycle in a manner consistent with a role in cell growth control [15]. Another TS gene, *p53*, may also regulate cell cycle progression and DNA replication [16].

Some cytokines are mitogenic for particular cell types and then also help stimulate their differentiation. Here the changes in gene expression must be even more complex: first, growth-promoting genes must be expressed, then later the proteins involved in differentiation must be induced. Examples are the CSFs that act on hematopoietic precursor cells, and IL-2 which is a growth factor for T lymphocytes. These are analyzed in more detail below.

4.2.2 Inhibition of cell growth by cytokines

Cytokines that inhibit growth might inhibit any of the mechanisms used by mitogenic growth factors for activating cell proliferation *(Figure 4.5)*. Cytokines such as IFNα/β, TGFβ or the TNFs that induce cytostasis might, for example, down-regulate growth-factor receptors, inhibit the signal transduction pathways involved in commitment or progression of cells through the cell cycle or inhibit expression of growth-factor-induced genes. Production of autocrine growth factors could also be impaired by cytostatic cytokines. Alternatively, increased expression of TS gene products or of their regulators (e.g. PK or phosphatases) might interfere with growth-promoting mechanisms.

Examples of some of these mechanisms are also known in cell culture systems. IFNα or β strongly inhibit proliferation of the Daudi Burkitt's lymphoma cell line, with associated rapid down-regulation of the *MYC* proto-oncogene [17] and another proto-oncogene, *FGR*, that codes for a tyrosine kinase possibly involved in signal transduction [18]. The mechanism appears to involve destabilization of the mRNAs, rather than blocks to transcription. At the same time IFN treatment induces other genes, and the cells cease to respond to autocrine B cell growth factors and take on a more

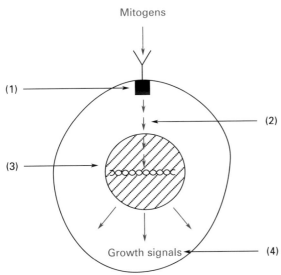

Figure 4.5: Possible mechanisms of growth inhibitory cytokines. Growth inhibitory cytokines may down-regulate or inactivate surface receptors (1), inhibit protein phosphorylation or other signal transduction events (2), down-regulate expression of essential genes (3), or alter the action of growth stimulatory proteins (4).

differentiated phenotype [19]. Hairy cell leukemia cells similarly lose sensitivity to B cell growth factors after IFN treatment [20]. It is not clear, however, whether the early changes in gene expression are sufficient to cause the later loss of growth potential and enhanced differentiation in these transformed B-cell systems. As indicated earlier, in other cell systems IFNs can act as negative growth regulators by down-regulating the EGF and other receptors. Thus the mitogenic potential of EGF for quiescent fibroblasts is reduced if the cells are pretreated with IFNs [21]. However, a combination of growth factors can overcome the IFN effect, perhaps by stimulating other signal transduction pathways, not sensitive to IFN.

In hairy cell leukemia IFNα induces the receptor for TNFα [22], perhaps rendering the cells sensitive to the growth inhibitory/cytotoxic effects of that cytokine.

As yet, there is little information on the relationship between growth inhibitors and TS genes. It has been suggested that the IFN genes may themselves act as TS genes (see Chapter 5). IFN treatment of Daudi cells modulates the phosphorylation state of the *RB1* gene product [23], but it is not clear whether this is a cause or an effect of the IFN-induced growth inhibition. Cells tend to accumulate in the G_1/G_0 phase of the cell cycle after IFN treatment and this may secondarily alter the balance of phosphorylation and dephosphorylation of the RB protein.

4.2.3 Cytokines in bone marrow

The great variety of cytokines involved in regulating hematopoiesis in bone marrow has been indicated in Chapter 1, where their individual properties were described. Hematopoiesis illustrates regulation of cell growth and differentiation by many different cytokines, sometimes acting synergistically, sometimes having antagonistic effects. As stated earlier, cytokines can induce production of other cytokines. IFNγ, TNFα and β, and IL-1 stimulate production of the CSFs that regulate hematopoietic cell differentiation *(Table 4.6)*. Monocytes/macrophages and activated T lymphocytes are major sources of these inducing cytokines. They also produce the CSFs themselves, allowing enhanced hematopoiesis during the response to foreign antigens *(Figure 4.6)*. Vascular endothelial cells and fibroblasts also produce CSFs such as GM-CSF and G-CSF, particularly in response to TNF or IL-1. Positive feedback may occur here

Table 4.6: Induction of colony stimulating factors by cytokines

CSF	Induction by			
	IFNγ	TNFα	TNFβ	IL-1
IL-3	+			
GM-CSF	+	+	+	+
G-CSF		+	+	+
M-CSF		+		

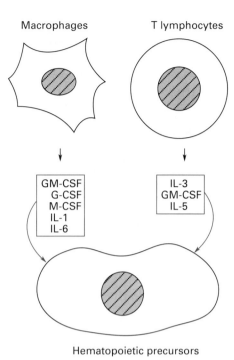

Figure 4.6: Stimulation of hematopoiesis by cells of the immune system. Normally stromal cells in bone marrow produce the CSFs which control growth and differentiation of hematopoietic precursor cells. During an immunological response to foreign antigens, activated macrophages and T cells can also produce CSFs and interleukins. Thus generation of mature blood cells is accelerated to help fight the infection.

because GM-CSF can induce TNF production in monocytes.

To balance these stimulatory effects, the actions of the CSFs on hematopoietic progenitor cells are limited by down-regulation of receptor levels in response to the CSFs themselves *(Table 4.4)*. There is a hierarchy of effects here, with some cytokines (e.g. IL-3) affecting several receptors and others down-regulating only their own and perhaps one other receptor type.

The physiology of control of hematopoiesis by CSFs and other cytokines will be covered in Chapter 5; here we consider the biochemical mechanisms of the stimulatory action on bone marrow progenitor cells. Compared to the knowledge of signal transduction mechanisms used by other cytokines, little is known about how CSFs exert their biological effects [24]. Probably different CSFs use different mechanisms since they often act synergistically *(Table 4.7)*. Bone marrow progenitor cell lines depend on the presence of these factors for viability, and CSFs stimulate basic cellular activities such as maintenance of ATP levels and uptake of sugar molecules. How this is achieved has not been demonstrated in detail, but the immediate cellular responses involve activation of several PKs. For example, IL-3 activates the serine/threonine-specific PK-C (see Chapter 3), apparently without inositol phospholipid hydrolysis; no changes occur in intracellular calcium metabolism, nor is cAMP involved in the intracellular signalling pathway. Tyrosine phosphorylation, however, probably is involved [24]. The M-CSF receptor, encoded by the *FMS* proto-oncogene, has intrinsic tyrosine kinase activity. Other CSF receptors (e.g. for IL-3) lack this activity but may be associated with kinases in the plasma membrane. Transfecting CSF-dependent cell lines with viral oncogenes coding for a tyrosine kinase can render further cell viability and proliferation independent of CSF. The functional relationship between enzymes of the PK-C and tyrosine kinase families is not yet clear, but

Table 4.7: *Synergistic interactions of colony stimulating factors*

CSF combinations	Synergistic effects
IL-3 + M-CSF	Increased percentage of hematopoietic progenitor cells in cell cycle
IL-3 + M-CSF + GM-CSF	Reduced concentration of CSF needed for stimulation of cell proliferation
IL-3 + M-CSF	Colony formation by monocytes and macrophages *in vitro*
GM-CSF + M-CSF	Colony formation by monocytes and macrophages *in vitro*
GM-CSF + IL-3	Megakaryocyte colony formation
GM-CSF + G-CSF	Growth of acute myeloid leukemia cells

cross-signalling can certainly occur. Activation of such pathways often leads to increased expression of proto-oncogenes coding for nuclear products (e.g. *MYC*), suggesting that these proteins help convey the proliferative signals to the DNA-synthesizing machinery. Consistent with this notion, cytokines such as IFNγ that impair proliferation of some hematopoietic cell types can inhibit expression of *MYC*. This, and probably additional lineage-specific genes, also play essential roles in the subsequent CSF-dependent differentiation of hematopoietic cells; however, we do not understand how cell proliferation and terminal differentiation (involving loss of ability to proliferate) are balanced during normal cell development. An important aspect is probably cross-regulation of genes coding for other cytokines that stimulate differentiation (e.g. the TNFs and IFNγ); these effects will be superimposed on the cross-modulation of CSF receptor levels described above.

4.3 Cytokines and the immune system

The numerous ways in which the immune system responds to invasion by foreign organisms or other antigenic agents depend critically on many cytokines. The effects of individual cytokines on particular activities of T and B lymphocytes and macrophages are described elsewhere; here we consider the interactions involved between the many factors and cell types, to present an overall picture of the importance of cytokines in immune regulation.

4.3.1 Interleukins and lymphocyte activation

Figure 4.7 summarizes the interactions between antigen-presenting cells and T cells that lead to an immune response. Antigens are taken up and processed by macrophages (and other cell types such as dendritic cells) and presented to resting T helper cells in association with class II MHC antigens on the macrophage cell surface. This complex interacts with the T-cell antigen receptor to trigger cell activation. At the same time the macrophages produce IL-1 and IL-6 which assist the processed foreign antigens to activate the T cells. Cell activation involves signal transduction pathways similar to those used by growth-factor receptors, including increased turnover of membrane phospholipids, elevation of cytoplasmic free calcium levels and activation of PK-C [25]. Several of these effects can be mimicked experimentally by exposing T cells to mitogenic proteins such as plant lectins or treating them with agents that elevate Ca^{2+} and stimulate PK-C. Additional signals, perhaps the result of IL-1 and IL-6 stimulation, are also needed for a full T cell proliferative response. Activated T cells also express increased numbers of IL-1 receptors. Other intracellular changes follow these initial activation events, culminating in DNA synthesis and cell division after 24–48 h. Continued stimulation and mitogenesis thus result in clonal expansion of an antigen-specific T cell population.

Cells become committed to a mitogenic response within 1–2 h of stimulation, but the molecular mechanism is not known. Among early effects of stimulation are expression of *FOS* (15 min), and of *MYC* (30 min) and IL-2 synthesis (within the first hour). The IL-2 receptor is also induced, allowing autocrine stimulation by IL-2. Subsequent changes, prior to DNA synthesis, include increased expression of transferrin receptors. Transferrin is necessary for iron uptake and for cell proliferation, but probably has no regulatory role in mitogenesis *per se*.

Synthesis of IL-2 and its receptor is almost certainly the key step in the response of T cells to antigenic stimulation. IL-2 gene expression requires various transcription factors, some of which may be activated by PK-C or by tyrosine-kinase mediated phosphorylation events induced by IL-1 or mediated by the T-cell antigen receptor. The T cell surface antigens CD4 and CD8 are believed to transduce signals during

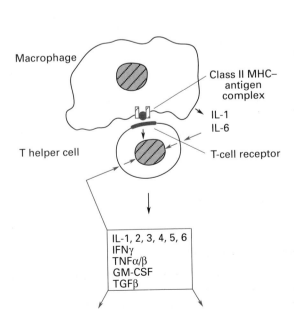

Figure 4.7: Activation of T lymphocytes in the immune response. Resting T helper cells are activated by processed foreign antigens combined with class II MHC antigens, presented on the surface of macrophages. The MHC–antigen complex is recognised by the T-cell antigen receptor which is linked to signal transduction pathways. The macrophages also produce IL-1 and IL-6 which independently activate T cells. The activated T cells produce many lymphokines and other effectors, some of which exert a feedback stimulatory effect leading to T-cell proliferation and differentiation; these cytokines also interact with many other target cell types to modulate their function.

antigen-mediated T-cell activation; they are physically associated with p56^{lck}, a member of the *SRC* family of tyrosine kinases and the product of the *LCK* oncogene. If CD4 molecules are induced to aggregate on the cell surface by anti-CD4 antibodies, p56^{lck} exhibits increased tyrosine kinase activity. Thus the p56^{lck}–CD4 complex can function similarly to a growth factor receptor with tyrosine kinase activity [26]. However, it is not yet clear what genes may be induced by this pathway. The subsequent binding of IL-2 to its high affinity receptor (the α/β dimer – see Chapter 3) does not raise cytoplasmic free Ca^{2+} levels or activate PK-C, but may stimulate a receptor-associated PK. Both IL-2 and IL-1 can induce *MYC* expression, although only IL-1 induces *FOS*. IL-2 also induces another proto-oncogene involved in cell cycle progression, *MYB*, with a time course similar to that of induction of the IL-2 receptor α sub-unit, peaking at about 8 h.

Among other new products synthesized by activated T cells are IFNγ, TNFα and β, IL-1, IL-3, IL-4, IL-5, IL-6, IL-7 and GM-CSF, several of which probably help promote the proliferation and differentiation of the T cells. For example, receptors for TNFs appear on T cells only after activation, suggesting that TNF responsiveness is involved in the differentiation of these cells (e.g. for acquisition of cytolytic activity by cytotoxic T cells). TGFβ is also produced by activated T cells and may inhibit growth and effector functions.

Just as T cells are activated by macrophage-mediated presentation of antigens, resting B lymphocytes are stimulated by interaction of antigens with surface immunoglobulin molecules [27]. This primes them to respond to cytokines produced by T cells, notably IL-1, IL-2, IL-4, IL-6, TNF and IFNγ *(Figure 4.8)*. As with T-cell activation the initial interaction of antigen with resting B cells triggers inositol phospholipid turnover, mobilization of intracellular calcium stores and activation of PK-C. These pathways presumably activate transcription of new genes, including those for the newly induced cytokines. There is some evidence that IL-4, and possibly other growth factors (e.g. human B cell growth factor II), can also activate resting B cells, perhaps via protein phosphorylation. Indeed, a full proliferative B-cell response requires stimulation of both antigen-dependent and IL-4-dependent pathways.

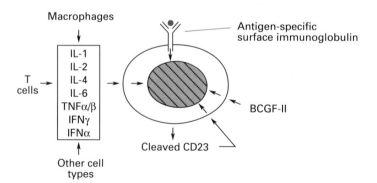

Figure 4.8: *Activation of B lymphocytes in the immune response. Small resting B cells are activated by antigens binding to immunoglobulins on the cell surface. This stimulates conversion to large proliferating B cells, which respond to T-cell- and macrophage-produced lymphokines as shown. These cytokines, together with B-cell specific growth factors such as BCGF-II and soluble CD23 molecules (cleaved from membrane-bound CD23 on the B-cell surface), stimulate cell proliferation and terminal differentiation into antibody-secreting plasma cells.*

B cells themselves produce specific autocrine factors that contribute to the proliferative response: the CD23 cell surface antigen is induced by mitogenic signals and then cleaved to give a soluble 25 kd protein which is a cell cycle progression factor.

The ultimate role of B cells is to secrete specific antibodies. Further cytokines control immunoglobulin isotype switching and terminal differentiation of proliferating B cells into antibody-secreting plasma cells [28]. IL-4 promotes isotype switching from IgM to IgG, and can also promote IgE production (see Chapter 5). Both these effects are antagonized by IFNγ. The terminal differentiation into plasma cells is controlled by IL-6, produced by many different cell types including activated macrophages and T cells. The IFNs may also influence this process.

As with T cells, negative regulatory cytokines probably also control B cell functions. Again TGFβ is a strong candidate for this role since it inhibits antibody production by B lymphocytes.

4.3.2 Other cytokines with immunoregulatory functions

As mentioned earlier, cytokines other than the T-cell specific factors influence the immune system. These effects are important because they allow the immune response to be amplified or damped down by products of other cell types that respond to quite different stimuli *in vivo*. *Table 4.8* lists some examples of cytokines and growth factors produced in response to tissue damage, inflammatory signals or systemic infections which can affect cells of the immune system [29].

TNFs are mitogens for activated T and B lymphocytes, resembling IL-1 in this respect. TNFα induces the α sub-unit of the IL-2 receptor (Tac antigen) on T cells. TNFs enhance proliferation of B cells and IL-1-induced secretion of antibody, and can stimulate expression of class I MHC antigens by lymphocytes. Many of the positive effects of the TNFs on immune functions are enhanced by IFNγ and, conversely, suppressed by TGFβ. TNFα is also stimulatory for many activities of monocytes and macrophages (e.g. secretion of IL-1 and IL-6 and expression of cytotoxic functions).

IFNs α and β can also influence immune cells. As with TNFs, induction of class

Table 4.8: Influence of cytokines on cells of the immune system

Cytokine	Effects on immune functions
IFNαs	Inhibition of LAK cell activity Stimulation of natural killer cell activity
TNFα/β	Stimulation of IL-2-mediated LAK cell activity
GM-CSF, TNFs, IL-1	Enhancement of IFNγ-stimulated cytotoxicity and phagocytosis by macrophages
TGFβ	Inhibition of T-lymphocyte activation by IL-2 Inhibition of pre-B-cell maturation Inhibition of natural killer cell activity

I MHC molecules is commonly observed, but only IFNγ induces class II MHC proteins. IFNα has been reported to promote maturation of cytotoxic T cells and can stimulate production of the immunoglobulin Fc receptor on macrophages, enhancing their phagocytic capacity. Another property of the IFNs, their stimulatory effects on natural killer lymphocytes, is described in Chapter 5.

References

1. Lyons, R.M. and Moses, H.L. (1990) *Eur. J. Biochem.*, **187**, 467.
2. Hughes, T.K. and Baron, S. (1987) *J. Biol. Regl. Homeost. Agents*, **1**, 29.
3. Piacibello, W., Lu, L., Williams, D. *et al.* (1986) *Blood*, **68**, 1339.
4. Zoon, K.C., Karasaki, Y., Zur Nedden, D.L., Hu, R. and Arnheiter, H. (1986) *Proc. Natl Acad. Sci. USA*, **83**, 8226.
5. Dexter, T.M., Coutinho, L.H., Spooncer,E., Heyworth, C.M., Daniel, C.P., Schiro, R., Chang, J. and Allen, T.D. (1990) *Ciba Found. Symp.*, **148**, 76.
6. Ensoli, B., Salahuddin, S.Z. and Gallo, R.C. (1989) *Cancer Cells*, **1**, 93.
7. Cross, F., Roberts, J. and Weintraub, H. (1989) *Ann. Rev. Cell Biol.*, **5**, 341.
8. Broek, D., Bartlett, R., Crawford, K. and Nurse, P. (1991) *Nature*, **349**, 388.
9. Cooper, G.M. (1990) *Oncogenes*. Jones and Bartlett, Boston.
10. Cross, M. and Dexter, T.M. (1991) *Cell*, **64**, 271.
11. Kouzarides, T. and Ziff, E. (1989) *Cancer Cells*, **1**, 71.
12. Bishop, J.M. (1991) *Cell*, **64**, 235.
13. Klein, G. (1990) *Tumor Suppressor Genes*. Marcel Dekker, New York.
14. Robbins, P.D., Horowitz, J.M. and Mulligan, R.C. (1990) *Nature*, **346**, 668.
15. Buchkovich, K., Duffy, L.A. and Harlow, E. (1989) *Cell*, **58**, 1097.
16. Sturzbecher, H.-W., Maimets, T., Chumakov, P. *et al.* (1990) *Oncogene*, **5**, 795.
17. Jonak, G.J. and Knight, E. (1984) *Proc. Natl. Acad Sci. USA*, **81**, 1747.
18. Sharp, N.A., Luscombe, M.J. and Clemens, M.J. (1989) *Oncogene*, **4**, 1043.
19. Exley, R., Nathan, P., Walker, L., Gordon, J. and Clemens, M.J. (1987) *Int.J.Cancer*, **40**, 53.
20. Paganelli, K.A., Evans, S.S., Han, T. and Ozer, H. (1986) *Blood*, **67**, 937.
21. Taylor-Papadimitriou, J., Shearer, M. and Rozengurt, E. (1981) *J. Interferon Res.*, **1**, 401.
22. Billard, C. and Wietzerbin, J. (1990) *Eur. J. Cancer*, **26**, 67.
23. Thomas, N.S.B., Burke, L.C., Bybee, A. and Linch, D.C. (1991) *Oncogene*, **6**, 317.
24. Dexter, T.M., Garland, J.M. and Testa, N.G. (1990). *Colony-Stimulating Factors*. Marcel Dekker, New York.
25. Berry, N. and Nishizuka, Y. (1990) *Eur. J. Biochem.*, **189**, 205.
26. Eiseman, E. and Bolen, J.B. (1990) *Cancer Cells*, **2**, 303.
27. Guy, G.R., Bee, N.S. and Peng, C.S. (1990) *Prog. Growth Factor Res.*, **2**, 45.
28. Finkelman, F.D., Holmes, J., Katona, I.M., *et al.* (1990) *Ann. Rev. Immunol.*, **8**, 303.
29. Arai, K., Lee, F., Miyajima, A., Miyatake, S., Arai, N. and Yokota, T. (1990) *Ann. Rev. Biochem.*, **59**, 783.

Further reading

Activation of lymphocytes

Altman, A., Coggeshall, K.M. and Mustelin, T. (1990) Molecular events mediating T cell activation. *Adv. Immunol.*, **48**, 227.

Cytokines in bone marrow

Kincade, P.W., Lee, G., Pietrangeli, C.E., Hayashi, S. and Gimble, J.M. (1989) Cells and molecules that regulate B lymphopoiesis in bone marrow. *Ann. Rev. Immunol.*, **7**, 111.

Cytokine networks

Opdenakker, G., Cabeza-Arvelaiz, Y. and Van Damme, J. (1989) Interaction of interferon with other cytokines. *Experientia*, **45**, 513.

Cytokine regulation of the immune response

Paul, W.E. (1989) Pleiotropy and redundancy: T cell-derived lymphokines in the immune response. *Cell*, **57**, 521.

Mechanisms of action of growth factors

Morrison, D.K. (1990) The Raf-1 kinase as a transducer of mitogenic signals. *Cancer Cells*, **2**, 377.

Mechanisms of action of growth inhibitory cytokines

Billiau, A. and Dijkmans, R. (1990) Interferon-gamma: mechanism of action and therapeutic potential. *Biochem. Pharmacol.*, **40**, 1433.

Mechanisms of action of hematopoietic growth factors

Zipori, D. (1990) Regulation of hemopoiesis by cytokines that restrict options for growth and differentiation. *Cancer Cells*, **2**, 205.

Oncogenes

Sukumar, S. (1990) An experimental analysis of cancer: role of *ras* oncogenes in multistep carcinogenesis. *Cancer Cells*, **2**, 199.

Tumor supressor genes

Marshall, C.J. (1991) Tumor suppressor genes. *Cell*, **64**, 313.

5
CYTOKINES IN HEALTH AND DISEASE

Since cytokines control an impressively diverse range of biological functions, it is not surprising that their correct production and action is necessary for good health. As a corollary, abnormal cytokine functions are implicated in many diseases. One or more cytokines are probably involved in every process where cell proliferation or differentiation occur, including tissue growth and turnover, generation of all types of blood cells, repair of wounds, and cellular responses to external changes. Diseases with abnormal growth patterns or inappropriate states of cell differentiation often include derangements of cytokine function; obvious examples include all types of cancer, non-malignant abnormalities of tissue growth, defects in healing or repair, and many kinds of hematopoietic deficiencies.

Another important function *in vivo*, involving many cytokines, is resistance to infections and invasions by foreign organisms. This can require both local activation of host defense mechanisms, as with wounds and tissue damage, and systemic responses such as those produced by virus infections. Several pathways can be mobilized, especially stimulation of immune responses, production of directly acting anti-viral agents (IFNs), and activation of inflammatory mechanisms.

5.1 Homeostatic regulation

In the healthy body the cytokines are largely concerned with maintaining the *status quo*, regulating the balance of cell numbers and activities between various tissues and cell types. As important homeostatic regulators, cytokines must be able to respond to changes in the extracellular environment, reflecting different nutritional circumstances, exercise or rest, or when particular demands are placed on certain tissues or cell types. The best understood of these homeostatic processes at present are the pathways controlling production of the various blood cell lineages – collectively known as hematopoiesis.

5.1.1 Control of hematopoiesis

Hematopoiesis needs to be controlled in order to maintain the correct levels of the various mature blood cell types [1]. This requires that the size of the pool of multipotential stem cells, from which all blood cells are derived (*Figure 1.3*), must be maintained, and commitment of these cells to differentiation along the possible lineages must be regulated to keep the correct balance between lineages. These needs pose complex problems of cellular co-ordination, especially since in adults all hematopoietic processes occur in one location – the bone marrow.

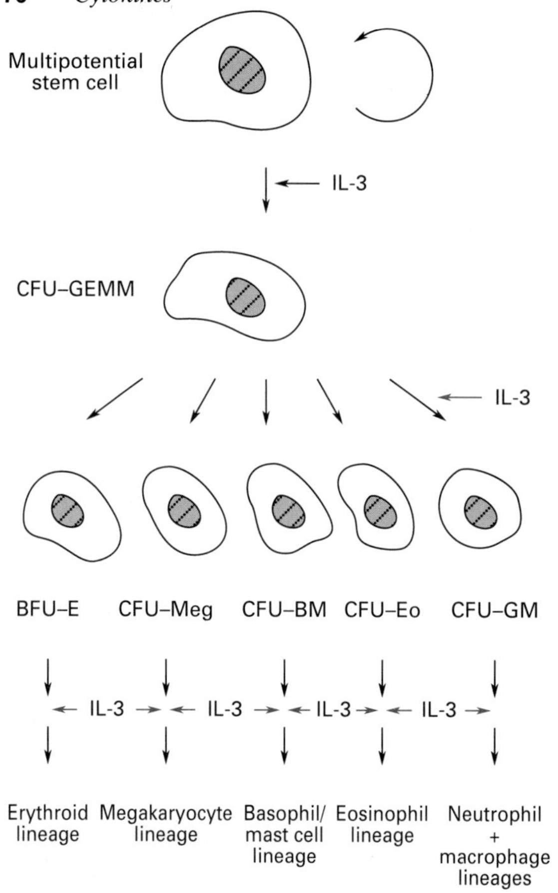

Figure 5.1: *Action of IL-3 in hematopoiesis. Self-renewing multipotential stem cells in bone marrow can differentiate into CFU-GEMM cells which then go on to develop into the immediate precursors of mature cells as shown. IL-3 induces proliferation and regulates differentiation at many stages in these pathways. BFU, burst-forming unit.*

Table 5.1: *Target cells for the colony stimulating factors*

Cytokine	Target cell types
IL-3	Monocyte precursors; granulocyte precursors; eosinophil precursors; megakaryocyte precursors; erythroid precursors (with EPO)
GM-CSF	Monocyte precursors; granulocyte precursors; eosinophil precursors; erythroid precursors (with EPO)
G-CSF	Granulocyte (neutrophil) precursors
M-CSF	Monocyte precursors
IL-5	Eosinophil precursors
EPO	Erythroid precursors

Several cytokines are known to be involved in control of hematopoiesis, and others may also be implicated (see Chapter 1). The colony stimulating factors GM-CSF, G-CSF, M-CSF and EPO are crucial, as are the interleukins IL-3 and IL-5. In addition, IL-1, IL-4 and IL-6 may have roles at various points along one or more differentiation pathways.

IL-3 acts very early in the process, when multipotential stem cells differentiate into CFU-GEMM cells (colony-forming units for granulocytes, erythrocytes, monocytes and megakaryocytes) which are precursors of the erythroid and myeloid lineages (*Figure 5.1*). The mechanism is unknown; IL-3 may change gene expression, allowing synthesis of products required for differentiation, or it may simply promote growth if stimulated cells are somehow more susceptible to other differentiation-inducing cytokines. IL-3 is known to act rather promiscuously as a growth stimulator during hematopoiesis [1] since it also promotes proliferation of later single-lineage progenitor cells (*Figure 5.1*).

Unlike IL-3, many other hematopoietic cytokines have more restricted types of target cells *(Table 5.1)*. EPO, G-CSF, M-CSF and IL-5 act mainly on precursors of erythroid cells, neutrophils, monocytes/macrophages and eosinophils, respectively. GM-CSF has intermediate specificity, stimulating growth and differentiation of CFU-GEMM, CFU-GM and monocytes themselves. IL-3 acts synergistically with some of the lineage-restricted cytokines in stimulating proliferation and differentiation of precursor cells. This suggests different roles (and mechanisms of action) for IL-3 and the CSFs; it also suggests how low levels of cytokines can control blood cell differentiation within the bone marrow, if one factor can enhance the actions of another. The hematopoietic progenitor cells in bone marrow are very sensitive to stimulation by these cytokines, even when the receptors are only partially occupied by ligand.

IL-1 also plays a role in hematopoiesis, enhancing the effects of IL-3 or M-CSF on multipotential stem cells or monocyte progenitor cells, respectively (an earlier name for IL-1α was hematopoietin 1). It probably acts in a permissive capacity, perhaps inducing receptors for other growth factors. It also induces synthesis of GM-CSF and M-CSF themselves. In accordance with these effects, IL-1 induces production of neutrophils *in vivo*. IL-6 also influences hematopoiesis, potentiating the effects of IL-3 and GM-CSF on cell proliferation, and, judging by its effects on certain cultured myeloid leukemia cell lines, may also promote myeloid cell differentiation. Rather similar roles have been suggested for another recently characterized cytokine, LIF [2].

Given the crucial importance of homeostatic regulation of hematopoiesis, it is not surprising that there are inhibitory as well as stimulatory cytokines *(Table 5.2)*. TGFβ is one example, acting particularly on marrow precursor cells in the earliest stages of the lineages [3]. Similarly, a recently described macrophage product, macrophage inflammatory protein-1α, slows proliferation of stem cells in bone marrow [4], while

Table 5.2: Stimulatory and inhibitory cytokines in hematopoiesis

Stimulatory cytokines	Inhibitory cytokines
IL-3	TGFβ
GM-CSF	Macrophage inflammatory protein-1α
G-CSF	TNFα
M-CSF	IFNα
EPO	IFNγ
IL-1, IL-4, IL-5, IL-6	

TNFα inhibits proliferation of myeloid precursor cells. IFNs inhibit proliferation of CFUs of all lineages, and some forms of aplastic anemia are associated with elevated production of IFNγ (and some IFNα) by a sub-set of activated T lymphocytes. IFNs can suppress the production of both erythroid and granulocyte/macrophage progenitor cells in bone marrow [5]. Note, however, that IFNγ can stimulate GM-CSF production by monocytes and T lymphocytes as well as IL-3 production by the latter; these effects will tend to counteract the depression of bone marrow function.

Since hematopoiesis occurs in the bone marrow of the adult, it is pertinent to ask where the many cytokines that influence this process are produced, and by what cell types. Probably hematopoiesis is normally under localized control within the marrow, the heterogeneous stromal cells acting as a source of paracrine stimulation or inhibition *(Figure 5.2)*. Progenitor cells lie in close contact with the bone marrow stroma, which needs to produce only cytokines that act over short distances; some (particularly IL-1α) may even act whilst still bound to the plasma membrane of the producer cells (see Chapter 2). Thus cell–cell contact may be very important in normal control of hematopoiesis [6]. When more dramatic or rapid responses are required, additional cytokine-mediated regulation can be brought to bear on the marrow from outside *(Figure 5.2)*. In response to acute infections, antigenic stimulation or inflammatory conditions, where increased hematopoietic activity is desirable in order to mount an attack on potentially pathological situations, cells of the immune system can produce the CSFs, IL-3 and IL-5. Indeed, erythropoiesis is always under such 'endocrine' control because the primary source of the essential cytokine EPO is the kidney. Production of EPO is closely regulated by the level of oxygenated erythrocytes in the blood, providing a sensitive mechanism for controlling erythroid cell development in response to the needs of the body.

Negative regulation of hematopoiesis is probably also mediated by both locally produced and peripherally delivered cytokines. For example, many cell types, includ-

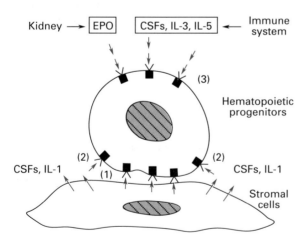

Figure 5.2: *Cytokine-mediated regulation of hematopoiesis. Proliferation and differentiation of hematopoietic progenitor cells is regulated by CSFs and other cytokines produced by stromal cells in the bone marrow. Some of these factors rely on direct cell–cell contact between producer and target cells (1), but localized actions of secreted cytokines are also likely (2). Erythropoiesis is controlled by EPO, produced predominantly by the kidneys. Immune stimulation or inflammatory responses cause T lymphocytes and macrophages to release further growth factors, which boost hematopoiesis above normal (3).*

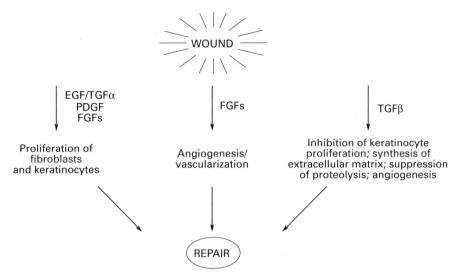

Figure 5.3: Mechanisms of wound healing. Various cells near a wound release cytokines that aid healing. Mitogenic factors (EGF, TGFα, PDGF and the FGFs) stimulate proliferation of fibroblasts and keratinocytes to replace damaged tissue. FGFs promote angiogenesis and vascularization leading to restoration of the blood supply to the repaired tissues. TGFβ has several important roles, as shown.

ing marrow cells, can synthesize TGFβ. However, when cells such as macrophages or platelets produce TGFβ (e.g. at sites of inflammation), the additional supply may temporarily suppress further proliferation and differentiation of these cell types in the marrow, limiting the inflammatory response. Thus bone marrow can be sensitively regulated in both positive and negative ways according to the needs of the body as a whole.

5.2 Wound healing

Several growth factors and other cytokines are involved in repairing tissue damage *(Figure 5.3)*, which is an important example of cytokine actions in health and disease.

5.2.1 The role of growth factors and inhibitory cytokines

Several growth factors that act on epithelial and fibroblastic cells are probably involved in wound healing *(Table 5.3)*, including EGF and TGFα (that both bind to the same receptor), PDGF and the FGFs. In the skin TGFα is both produced by, and acts upon,

Table 5.3: Growth factors and cytokines involved in wound healing

---
EGF/TGFα
PDGF
acidic FGF
basic FGF
TGFβ
---

keratinocytes and macrophages. Its mitogenic effect on keratinocytes would lead to proliferation of these cells near a wound, particularly if the site had been infiltrated by macrophages that produced the factor locally. FGFs may also contribute to cell proliferation (e.g. of fibroblasts) at a wound, but in addition, FGFs promote division of vascular endothelial cells and stimulate angiogenesis, ensuring the necessary vascularization of newly repaired tissues.

An additional role in wound healing has been attributed to TGFβ. Blood platelets are a major source of this cytokine at sites of tissue damage. TGFβ is a potent growth inhibitor for many cell types (including keratinocytes), so it may antagonize mitogens involved in wound repair; but it can also promote biochemical events associated with laying down new connective tissue, such as synthesis of extracellular matrix components (collagen, fibronectin and proteoglycans). TGFβ also stimulates cells to increase production of plasminogen activator inhibitor-1, tissue-specific inhibitor of metalloprotease, and perhaps other inhibitors of proteolytic enzymes. Suppressing proteolytic activities around wounds would assist formation of new connective tissue. Interestingly, TGFβ may itself be activated from its precursor form (see Chapter 2) by proteases such as plasmin that are initially present at the site of a wound. TGFβ is also angiogenic and can act as a chemotactic agent, attracting other cell types involved in wound healing such as monocytes and fibroblasts. In view of these pleiotropic effects it is not surprising that TGFβ does indeed accelerate wound healing and enhance the strength of the wound repair in experimental animals [7].

The balance of positive and inhibitory effects in wound healing provides a good example of how the different cytokines interact to produce the correct net effect *in vivo*. No doubt these and other factors similarly regulate growth and turnover of many other tissues and cell types. Given the fine balance required to produce healthy tissues it is not surprising that abnormal cytokine activity is seen in many diseases involving disruption of growth control.

5.3 The immune system

As shown in Chapter 4, numerous factors influence lymphocyte functions, and so there are many points where disrupted regulatory loops can lead to diseases. Several autoimmune diseases stem from inappropriate immune responses, particularly to self-antigens, and they involve a number of cytokines. The immunodeficiency syndromes also constitute a class of diseases associated with abnormal cytokine regulation.

5.3.1 Cytokines and autoimmune diseases

When antigens presented to the immune system derive from the organism's own tissues, rather than a foreign cell, self-destruction and autoimmune disease can follow. Several cytokines have been implicated in this process *(Table 5.4)*, particularly IFNs and TNFs. The initial causes of autoimmune diseases are often unknown, although infectious organisms probably help sensitize the immune system to self-antigens. Subsequently, over-production of cytokines could contribute to the chronic phase. For example, in systemic lupus erythematosus (SLE), an apparently acid-labile form of IFNα appears in the circulation, which may enhance expression of histocompatibility antigens, leading to activation of immune responses. The IFN may also stimulate B lymphocytes to produce auto-antibodies. Production of TNFα may cause glomerular

Table 5.4: Cytokines implicated in autoimmune diseases

IFNα (including an acid-labile species)
IFNγ
TNFα
TNFβ
IL-1β

nephritis leading to the kidney damage associated with SLE. Interestingly, the genes for both TNFα and β are closely linked to the MHC genes on human chromosome 6 and there is a strong connection between an individual's class II MHC phenotype and susceptibility to nephritis in SLE. Likewise, the class II MHC phenotype affects susceptibility to celiac disease, in which autoimmune destruction of jejunal enterocytes may involve locally produced TNFβ.

Perhaps virally induced insulin-dependent diabetes mellitus arises when infecting viruses, such as certain strains of Coxsackie B, cause production of IFNα by pancreatic β cells, which then leads to increased expression of MHC class I antigens on these cells, activating inappropriate immune responses that contribute to the destruction of the insulin-producing β cells [8]. There may also be a secondary role for IFNγ produced by T lymphocytes as part of the immune response; this IFN can activate production of TNFα and IL-1β by monocytes as well as introducing class II MHC expression.

IFNα may posssibly be involved in multiple sclerosis, again by inducing class II MHC antigen expression, in this case on astrocytes or Schwann cells in the brain, resulting in increased presentation of tissue-specific auto-antigens to T lymphocytes. Subsequent attack by cytotoxic cells may cause the demyelination that leads to the loss of co-ordination and paralysis typical of this debilitating disease.

5.3.2 Cytokines and AIDS

The human immunodeficiency virus (HIV) that causes acquired immunodeficiency syndrome (AIDS) infects and destroys a sub-set of T lymphocytes (CD4-positive T helper cells). These are an important source of many cytokines, including IL-2, IFNγ and CSFs, and their disappearance probably explains the reduced levels of these cytokines in AIDS patients. Mitogen-stimulated cultured lymphocytes from some HIV-positive individuals show deficient production of IFNγ, and the extent of deficiency has prognostic value for the risk of developing symptoms of AIDS. Such cytokine deficiencies, amongst other changes, may underlie the multiple problems of AIDS, including opportunistic infections, development of tumors and autoimmune damage to nervous tissue. AIDS patients show reductions in several cytokine-regulated defensive functions *(Table 5.5)*. The reduced immunosurveillance typical of AIDS

Table 5.5: Defective defense mechanisms in AIDS

Reduced production of IL-2, IFNγ and CSFs by T helper cells
Impaired cytotoxic T-cell activity
Reduced natural killer cell activity
Decreased macrophage activity
Impaired antigenic stimulation of B lymphocytes

probably allows uncontrolled outgrowth of tumors such as the otherwise rare Kaposi's sarcoma and lymphomas arising from clonal proliferation of B cells infected with Epstein–Barr virus.

Peripheral blood cells from AIDS patients are defective in virally induced production of normal IFNα, although the basis for this is not known. In contrast, an unusual form of IFNα appears in the circulation of HIV-positive individuals before they progress to the clinical stages of AIDS-related complex and AIDS [9]. This IFNα probably differs structurally from the usual IFNα species since it is (atypically) acid-labile. The stimulus for its production is not known; it may not be the HIV itself. This IFN fails to protect AIDS patients against opportunistic viral infections; rather, its presence may be harmful, leading to autoimmune diseases. A similar IFN is found in other disorders with autoimmunity, such as SLE. The dementia seen at late stages in AIDS is believed to result from autoimmune destruction of brain tissue; IFN produced by inflammatory macrophages could also contribute to neurological symptoms. The high circulating levels of acid-labile IFNα may also be responsible for systemic symptoms of AIDS (fever and tiredness) and hematopoietic deficiencies. Thus excessive and deficient cytokine production can be equally deleterious, each causing an imbalance in processes that are normally tightly regulated.

IFNα and IFNγ both inhibit replication of the HIV in lymphocytes and monocytes *in vitro*. IFNγ inhibits the reverse transcriptase of HIV, an effect enhanced by GM-CSF. Since this enzyme is essential for HIV replication, the deficiency of normal IFN production in AIDS patients may be a necessary condition for the virus to spread to further cells and for the disease to progress. Clearly the acid-labile IFNα does not prevent HIV replication, but this may reflect the progressive failure of host defense mechanisms in a self-perpetuating downward spiral which underlies the features of AIDS.

5.4 Control of virus infections

The body combats invading viruses in many ways, including immune and non-immune pathways, and cytokines are essential to co-ordinate these effects. Viral particles are inactivated by neutralizing antibodies secreted by plasma cells; granulocytes and macrophages engulf micro-organisms by phagocytosis and can also release agents that destroy virus-infected cells; cytotoxic T lymphocytes kill cells by recognizing foreign antigens on cell surfaces; and IFNs directly induce a state of resistance to viral replication in a wide variety of target cells. It is this last aspect of the body's anti-viral defenses that will now be discussed.

5.4.1 The role of IFNs

In contrast to IFNγ which is induced by mitogenic or antigenic stimulation of T lymphocytes, IFNα and β can be induced directly by viruses following infection of host cells. Viruses also induce other cytokines, particularly TNFs and IL-2, but these do not themselves induce an anti-viral state in cells. Sequences upstream of the IFN genes (see Chapter 2) switch on transcription in response to viruses [10]. Many viruses produce double-stranded (ds) RNA as replication intermediates, and dsRNA can induce IFN gene transcription. However, dsRNA cannot be the only natural inducer of IFN production since IFNs can be induced by other cytokines such as the TNFs. Production of IFNs by virus-infected cells is only transient, mainly because the IFN genes rapidly lose the ability to be transcribed. The physiological purpose and biochemical mechanism of this phenomenon are not clear, but continuous production of IFNs over a long period could be harmful to the body. In any case, transient production of IFNs and development of an anti-viral state even for just a few hours may suffice to prevent the spread of viruses that must replicate rapidly and cannot remain dormant in infected cells.

Secreted IFNs appear in the bloodstream and other body fluids, but they may also act locally in a paracrine or autocrine fashion, decreasing the probability that a virus infection will spread from cell to cell within a tissue. Local concentrations of these cytokines may be quite high for transient periods to achieve this effect.

Unlike the immune system, where antigen specificity is paramount, IFNs produced in response to one virus are equally capable of protecting cells against quite unrelated viruses [11], so the mechanisms of action must be non-specific. Development of an anti-viral state in cells exposed to IFNs requires new gene expression, and the means by which IFN–receptor interactions switch on these new genes is currently a subject of intense research. Production of new viral particles in an infected cell requires protein synthesis, and among the novel proteins synthesized by IFN-treated cells [12] are two enzymes that can regulate protein synthesis. *Figure 5.4* illustrates the action of these 'anti-viral proteins'. The unusual enzyme 2'5'-oligoadenylate synthetase (actually a small family of enzymes) uses ATP to synthesize short oligomers of up to 12 adenylate residues linked by 2'5' phosphodiester bonds. These $2'5'(A)_n$ molecules allosterically activate a latent ribonuclease, RNase L, that degrades viral and cellular mRNA and ribosomal RNA. The other enzyme is a PK which phosphorylates initiation factor eIF-2, a protein required for all polypeptide chain initiation. Phosphorylated eIF-2

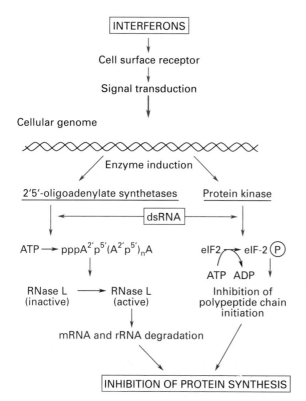

Figure 5.4: Mechanisms of action of IFN-induced anti-viral pathways. The anti-viral state induced by IFNs depends partly on two enzymes, 2'5'-oligoadenylate synthetase and a PK that specifically phosphorylates the protein synthesis initiation factor eIF-2. Both enzymes are activated by dsRNA and when active inhibit both viral and cellular protein synthesis.

inhibits initiation of viral and cellular protein synthesis. Both 2'5'-oligoadenylate synthetases and the PK require dsRNA for their activation. Since dsRNA is produced during the replication of many (but not all) viruses it may provide the signal both for induction of IFNs and for activation of the anti-viral enzymes induced by IFNs. Small RNA viruses of the picornavirus group, such as poliovirus, are particularly susceptible to inhibition by the dsRNA-activated mechanisms. It may seem surprising that the $2'5'(A)_n$- and PK-mediated pathways inhibit viral and cellular protein synthesis indiscriminately. However, generalized inhibition of protein synthesis in infected cells, even if it leads to cell death, may be tolerable to the organism as a whole provided it has the desired effect of preventing the spread of infection. Alternatively, the host cell may be able to withstand a period of reduced protein synthesis more readily than a virus, and may recover its ability to make new proteins after a few hours. It is also possible that the inhibition of translation, though non-specific, is localized within the cell and restricted to the close vicinity of replicating viral particles, where dsRNA is being produced.

IFNs can induce anti-viral activity in their target cells by other mechanisms in addition to those described above. An IFN-induced protein called Mx is particularly involved in inhibition of influenza virus replication [13]. The mechanism involves inhibition of viral transcription, although the details are not yet clear. IFN can also block other stages in the life cycles of various viruses, including initial penetration into the cell by endocytosis (e.g. in the case of vesicular stomatitis virus), and the final assembly and budding of viral particles from the cell surface (in retrovirus infections). Furthermore, IFN enhances the activity of natural killer cells against virus-infected cells [14].

Induction of these multiple responses by IFN probably allows the body to resist a broad spectrum of viruses with different mechanisms of replication – indeed all viruses show some degree of sensitivity to IFNs. The IFNs therefore play an important role in fighting many kinds of viral infection, and elevation of IFN levels in body fluids is usual during the acute stage of viral disease. Nevertheless, a number of viruses have evolved strategies for evading some of the IFN-induced mechanisms *(Table 5.6)* and these effects may partly explain why our defenses against viruses are less than perfect [15]. These strategies are probably equally important to acutely infectious viruses and to those such as herpes viruses, Epstein–Barr virus or retroviruses like HIV that

Table 5.6: Viral mechanisms for evasion of IFN-induced anti-viral pathways

Pathway	Virus	Mechanism of inhibition
$2'5'(A)_n$/RNase L	Vaccinia	Inhibition of RNase L
	Herpes simplex	Inhibition of RNase L
	EMC[a] (in mice)	Inhibition of RNase L
	Reovirus	Binding of dsRNA activator
dsRNA-activated PK	Adenovirus	Inhibition by small viral RNA
	Epstein–Barr	Inhibition by small viral RNA
	Vaccinia	Inhibition by induced protein
	Influenza	Inhibition by induced protein
	HIV-1	Down-regulation of PK
	Poliovirus	Degradation of PK
	Reovirus	Binding of dsRNA activator

[a]Encephalomyocarditis virus

establish persistent or latent infections in tissues, since they must all evade the anti-viral response during the initial infection. With persistent infections, preventing long-term pathological effects will depend on the efficacy of subsequent immune and other cell-mediated anti-viral defenses.

Some other cytokines may play additional roles in mediating direct anti-viral effects (i.e. not involving the immune system). TNFs can induce anti-viral activity in some situations, although this may be a result of TNF-induced IFN synthesis. IFNγ is certainly an anti-viral agent, although it acts more slowly than the α or β IFNs. However, the claimed anti-viral role of IL-6 (which led some laboratories to call it IFNβ2) is no longer accepted. It is important to emphasize that any one cytokine will have different effects on different cell types, as well as on different viruses. For example, TNFα can probably *induce* viral replication in HIV-infected lymphocytes by activating the transcription factor NF-κB which enhances HIV gene expression. However, HIV replication in T lymphocytes is inhibited by IFNα or β, and in monocytes it may be impaired by IFNγ acting in concert with GM-CSF [16].

It is useful to look at the IFN-mediated anti-viral effects as part of the overall set of anti-viral defense mechanisms. The rapid response of IFN synthesis (within a few hours of infection) will be the first defense. This will limit early viral replication and also cause some early symptoms of infection such as fever (IFNα has pyrogenic activity *in vivo*). In the longer term, however, induction of MHC-dependent cell-mediated and humoral immunity, and killing and phagocytosis of virus-infected cells and their debris become more important. Particularly for prophylaxis against future infections, the ability of the immune system to 'remember' earlier exposure to foreign antigens is crucial. There is no equivalent 'memory' within the IFN-mediated anti-viral pathways.

5.5 Cytokines and cancer

Since cancers are diseases of disordered cell growth and differentiation, it is not surprising that cytokines are implicated in carcinogenesis. A great variety of events can

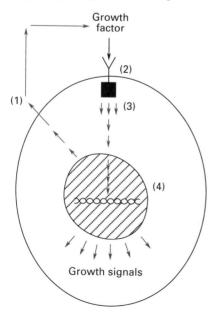

Figure 5.5: Molecular mechanisms of carcinogenesis. Inappropriate expression of autocrine growth factors (1) or their receptors (2), uncontrolled signalling by mutant receptor molecules or pathways (3), or permanent activation of enzymes or gene expression within the cell (4), could all lead to excessive proliferation and/or invasiveness by transformed cells.

go awry in the development of a cancer cell, reflecting the many steps by which cytokines mediate their effects on target cells *(Figure 5.5)*. For example, an important stage in the multi-step pathways leading to some leukemias may be dysregulation of the action of growth-stimulatory interleukins of CSFs in the bone marrow. Similar consequences might follow absence of a growth-inhibitory cytokine or failure of a cell to respond to it.

Cytokines are also involved in producing some of the symptoms of advanced cancer that are often the cause of death. The dramatic loss of body weight and negative nitrogen balance of cancer cachexia, and the degradation of bone (osteolysis) seen in many patients, are at least partially due to elevated levels of some cytokines.

5.5.1 Cancers as diseases of disordered signalling

There is no one single cause of cancer. Cancers are a very heterogeneous group of diseases and the molecular mechanisms of tumorigenic transformation may be quite distinct in different cell types. Moreover, conversion of a normal cell into a fully fledged cancer cell involves multiple steps, some perhaps genetically determined and some the result of environmental influences. The sequence of phenotypic changes may differ from one example to another. As a generalization, however, we can regard cancers as diseases in which cellular signalling mechanisms have been disrupted so as to remove growth and behavior of the cell from normal constraints *(Table 5.7)*. This could mean permanent activation of a receptor to produce growth-stimulatory signals whether or not its cytokine ligand is bound. Abnormal activity of a plasma-membrane bound tyrosine kinase, a G protein or an enzyme such as phospholipase C (see Chapter 3) could lead to over-stimulation of growth-promoting pathways. Inappropriate expression of PKs or phosphatases downstream of these initial signal transduction events could have similar effects. Other stages in the multi-step carcinogenic process could include over-expression of genes that are normally switched on only transiently by cytokine stimulation, or mutation of such genes to produce abnormal products. Some of the genes involved have been identified as proto-oncogenes – that is, genes that

Table 5.7: Possible mechanisms of disordered cell signalling in cancer

Changes in signal transduction

Mutation of growth factor receptors to a ligand-independent state, leading to excessive production of growth-stimulatory second messengers

Mutation of receptor-associated proteins such as G proteins or tyrosine kinases of the *SRC* family, with the same consequences as above

Mutation or over-production of downstream components in signalling pathways (e.g. PKs or phosphatases)

Changes in gene regulation

Inappropriate expression of growth-promoting genes (in the absence of the normal regulatory signals)

Mutations in growth-promoting genes, resulting in over-active products

Impaired expression of growth-inhibitory genes

Mutations in growth-inhibitory genes, leading to impaired activities of their products fulfil normal roles in untransformed cells but which can contribute to growth transformation when they function aberrantly [17]. *Table 5.8* shows that there is a close relationship between some of the proto-oncogenes and the components of cytokine-activated pathways. For example, the *SIS* gene codes for the B chain of PDGF, and the *INT2* gene encodes an FGF-like protein; *ERBB* and *FMS* encode the EGF and M-CSF receptors, respectively. Several proto-oncogene products are tyrosine kinases, perhaps similar to the growth factor receptor tyrosine kinases, while the *RAS* gene family is related to the α sub-units of G proteins. Products of other proto-oncogenes (e.g. *FOS* and *JUN*) are transcription factors that probably control the expression of other growth-regulating genes. The number of known proto-oncogenes is still increasing as further components of growth-promoting pathways are identified.

More recently, genes have been identified which are believed to play an opposite role in growth control [18]; these are the TS genes *(Table 5.9)*. Loss or inactivation of TS genes, particularly of both copies of a gene, may be associated with cell transformation. Examples are the *RB1* gene, loss of which is associated with hereditary retinoblastoma, and *p53*, the product of which helps regulate DNA replication. So far no TS gene has been directly linked with a cytokine or its receptor, but there are suggestions that genes for IFNα and β are deleted or translocated from human chromosome 9 to another site (which may inactivate them) in some cases of acute

Table 5.8: Classes of (proto-)oncogenes and their relationships to growth-regulatory factors

Oncogene	Normal cellular product
Oncogenes coding for growth factor-like products	
SIS	PDGF B chain
INT2	FGFs
HST	FGFs
FGF5	FGFs
Oncogenes coding for receptor-like products	
ERBB	EGF receptor
ERBB2	Related to EGF receptor
FMS	M-CSF receptor
KIT	Related to M-CSF receptor
Oncogenes coding for proteins involved in signal transduction	
SRC	
LCK	
FPS/FES	
RAS family	Related to G proteins
RAF	PKs
Oncogenes coding for nuclear proteins	
MYC	
FOS	Transcription factors
JUN	Transcription factors
MYB	Transcription factors

Table 5.9: Possible roles of tumor suppressor gene products

Down-regulation of growth factor receptors
Inhibition of growth-stimulatory signal transduction pathways
Repression of growth-stimulatory gene expression
Interference with processes required for DNA replication
Induction of growth-inhibitory cytokines
Stimulation of terminal cell differentiation

lymphoblastic leukemia or non-Hodgkin's lymphoma [19]. This may also occur in other malignancies. As the list of TS genes grows, more components of cytokine signalling pathways that are inhibitory for cell proliferation or needed for terminal differentiation will probably be identified.

The leukemias provide examples of malignant disease associated with disorders in cell signalling *(Table 5.10)* [20]. In some myeloid leukemias, inappropriate activation of the IL-3 gene in immature myeloid cells may provide an autocrine stimulus for cell proliferation without differentiation. Production of IL-3 by cells that also respond to it removes them from their normal dependence on T-cell produced IL-3. Unregulated endogenous expression of GM-CSF might have a similar effect. Some patients with acute myeloid leukemia do show abnormal GM-CSF production, although experiments in which this gene was constitutively expressed in transgenic mice failed to produce any tumors. In some cases of myeloid leukemias associated with previous cytotoxic drug treatment, one copy of the region of chromosome 5 containing the IL-3 and GM-CSF genes (as well as genes for other cytokines and receptors – see Chapter 2) is lost, but the relevance of this to the malignant phenotype of the cells is not known.

IL-2 may also be involved in leukemogenesis. In an adult T-cell leukemia caused by the human retrovirus HTLV-1, which infects T helper cells, there is constitutive expression of the IL-2 receptor [21], instead of the normal regulation by antigenic stimulation. Such cells could proliferate in an autocrine or paracrine response to IL-2.

Table 5.10: Putative disorders of cellular signalling in leukemias

Leukemia type	Possible molecular basis
Acute myeloid leukemia	Autocrine stimulation by inappropriately expressed IL-3 or GM-CSF
	Loss of part of chromosome 5 carrying genes for several hematopoietic growth factors (perhaps allowing action of mutant forms of these factors expressed from the homologous chromosome)
Adult T-cell leukemia	Constitutive expression of IL-2 receptor and/or loss of dependence on IL-2 for proliferation
Non-Hodgkin's T-cell leukemia	Autocrine stimulation by IL-2
Human T-cell leukemia; virus-induced leukemia	Autocrine stimulation by IL-4
Lymphomas induced by Epstein–Barr virus; myelomas	Autocrine stimulation by IL-6.

As the IL-2-driven clone proliferates, a further genetic change might bypass the requirement for IL-2 altogether, resulting in the completely autonomous proliferation that is typical of adult T-cell leukemia. Some non-Hodgkin's lymphoma cells may also proliferate uncontrollably because of an autocrine IL-2/IL-2–receptor stimulatory loop.

Other growth-stimulatory interleukins are also implicated in development of leukemic phenotypes. IL-4 is produced by some transformed T-cell lines and HTLV-1-infected T cells in culture; the latter possess IL-4 receptors, indicating that they could be susceptible to autocrine stimulation by this mitogenic cytokine. Likewise, IL-6 is a growth factor for myelomas and B lymphoblastoid cell lines immortalized by Epstein–Barr virus [22]. Thus in principle, any cytokine or receptor that has a normal role in promoting hematopoietic cell growth and division could be a growth stimulus to leukemic cells when genetic changes remove normal constraints on proliferation.

5.5.2 The balance between growth stimulatory and growth inhibitory cytokines

Normally several external factors control the ability of cells to proliferate and differentiate, including growth factors and other mitogenic cytokines, and growth inhibitory agents that limit the proliferation of cells and/or stimulate them into terminal differentiation. For a particular cell type, the kinds of cytokine receptors present on its surface will largely determine the factors to which it responds. Thus imbalances in the levels of positive and negative cytokines or changes in the cytokine receptors on the plasma membrane can lead to loss of growth regulation. Such mechanisms, whilst not necessarily sufficient to cause a cell to become malignant, could certainly contribute to transformation as one of the many steps involved in the multi-stage pathway of carcinogenesis.

Increased growth factor concentrations may occur locally if a transformed cell starts producing such molecules, and if the relevant receptor is present on the same cells an autocrine stimulatory loop will be set up. TGFα is produced by a number of carcinomas and transformed fibroblasts, and acts on cells via the same receptor as EGF. This receptor is closely related to the product of the v-*erb B* viral oncogene which can transform cells to a tumorigenic phenotype, probably by expressing large amounts of a permanently activated form of the receptor [23]. These observations provide further evidence that the EGF/TGFα group of cytokines are important in regulating the behavior of some tumor cells. Similarly, IGF-I is both produced by and acts upon breast carcinoma cells to promote their proliferation; IGF-I synthesis and action are promoted by estrogens in breast and uterine epithelial cells, providing one mechanism by which these steroids exert a growth-promoting effect on their target cells [24]. The FGFs too may play a role in carcinogenesis; they can be produced by some sarcomas and, in a mouse model, a related protein is induced when mouse mammary tumor virus DNA inserts near the *INT2* proto-oncogene [24]. *INT2*, or genes closely related to it, appears to be expressed in several human tumors, including Kaposi's sarcoma which often occurs in AIDS patients. Interestingly, growth factors of the FGF family may also promote angiogenesis, and when this occurs in a tumor the improved blood supply will help the tumor cells survive and proliferate. This is an example of tumor-derived cytokines acting in a paracrine manner on the normal cells around the tumor to produce effects such as vascularization and tissue hyperplasia *(Figure 5.6)*.

The relationship between normal growth factors and proto-oncogene products

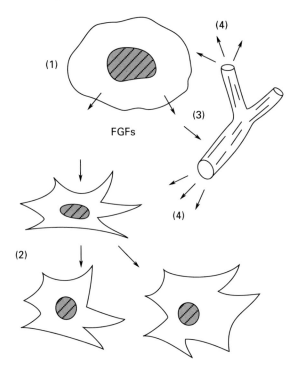

Figure 5.6: Paracrine effects of tumor-derived cytokines. Cytokines produced by tumor cells can alter the activities of surrounding normal cells. FGFs secreted by tumors (1) can cause proliferation of adjacent normal fibroblasts (2), leading to local hyperplasia. FGFs also stimulate angiogenesis (3), thus improving the delivery of nutrients and oxygen to the tumor and surrounding tissues (4).

was first revealed by the discovery that the PDGF B chain (see Chapters 1 and 2) is encoded by the *SIS* proto-oncogene. In the simian sarcoma virus a tumorigenic form of this gene, v-*sis*, occurs and is responsible for acute transformation of cells and production of tumors by this virus. PDGFs and related molecules stimulate early events associated with entry of quiescent cells into the cell cycle, which promote growth or lead to cell transformation. Cellular responses to PDGF (probably involving changes in gene expression) commit the cells to proliferation; further changes elicited by the growth factor (including induction of a transport system for getting glucose into the cell) may then drive the cells through the division cycle. Other growth factors can also induce this glucose transporter, which may be necessary to sustain the high rate of carbohydrate metabolism characteristic of tumor cells.

A fine balance must be struck between cell proliferation (e.g. for tissue replacement or wound repair) and restriction of inappropriate cell growth. When this balance is disrupted in favor of proliferation, tumors can develop. Indeed, cancers could be seen as 'wounds that never heal' although this is certainly an over-simplification. Restriction of inappropriate cell growth depends on the growth-inhibitory cytokines, prominent among which are IFNs and TGFβ. Other agents, such as the TNFs and IL-1, can also inhibit cell proliferation, although they are mitogenic for some cell types. As yet there is little direct evidence of tumorigenesis caused by non-production of, or non-responsiveness to, growth-inhibitory cytokines. A possible example is the frequent loss of IFN genes associated with abnormalities of human chromosome 9 in several neoplasias mentioned earlier. The TS genes, whose products may be necessary for normal growth regulation and loss of which may contribute to the transformed state, are less well characterized than the oncogenes.

Growth-inhibitory cytokines could control proliferation of normal cells or impair growth of tumors *in vivo* by many mechanisms (*Figure 4.5*). Direct effects on the behavior of a cell are probably responsible in many situations. For example, the IFNs can down-regulate expression of cellular proto-oncogenes such as *MYC* in some cell types. Enzymes associated with DNA replication may also be targets for down-regulation. Inhibition of protein synthesis by the same pathways that are used to induce the antiviral state (see Section 5.4) may impair the ability of a cell to grow and divide. IFNγ inhibits growth of some carcinoma cell lines by inducing an enzyme, indoleamine-2,3-dioxygenase, that degrades tryptophan and starves the cell of this essential amino acid. IFNs can also probably down-regulate receptors for some growth factors such as EGF, insulin or the essential iron-delivery protein transferrin [25], and IFN treatment of cells inhibits mitogenesis in response to growth factors [26]. *In vivo*, additional IFN-regulated mechanisms may be very important in controlling abnormal cell proliferation, since cytotoxic cells such as natural killer lymphocytes and cytotoxic T cells are activated or matured by IFNs, enabling them to attack tumor cells. IFNγ also stimulates macrophages.

Although TGFβ can act as a growth factor (hence its name), it can block the mitogenic effects of EGF, PDGF, the FGFs and insulin, and inhibition of cell growth is probably its major role *in vivo*. Many normal and transformed cells both produce and respond to TGFβ, so a negative autocrine mechanism may restrict their proliferation. The loss of TGFβ production, or of its receptor, could thus remove at least one limitation on cell proliferation. This may happen during progression of a tumor to a more malignant form; for example, in carcinoma of the breast TGFβ production is observed but the cells can lose their ability to respond. The mechanism of action of TGFβ is not clear but it has been shown to inhibit induction by EGF of proteolytic enzymes, such as stromelysin, which are involved in destruction of extracellular matrix. Extracellular proteolysis may be necessary before highly metastatic carcinomas can invade surrounding tissues. Indeed, TGFβ is also able to induce synthesis of connective tissue components such as collagen, counteracting the effects of the extracellular proteases.

TNFs can inhibit tumor cell growth by direct cytotoxic effect, at least in some experimental systems [27]. In other cases no cell death is observed but the cells cease to proliferate and may even differentiate. The cytotoxic effects of the TNFs appear to involve production of arachidonic acid (see Chapter 3), further metabolism of which can generate highly reactive oxygen free radicals that are extremely damaging to DNA. In susceptible cells the DNA is fragmented. *In vivo,* TNFα is also implicated in the ability of natural killer cells or macrophages to destroy tumor cells, perhaps by local TNFα production on the cell surface. TNFα can also act in conjunction with IL-2 to activate another class of lymphocytes, lymphokine-activated killer (LAK) cells, which are believed to be important in non-antigen-specific killing of tumor cells *in vivo*. Such effects suggest applications of cytokines for cancer therapy (see Chapter 6).

Some of the inhibitory effects of IL-I may also be mediated by TNF (which is induced by IL-1), although different cells respond to TNF and IL-1. Since many of the cytokines induce each other (see Chapter 4), dissecting the contributions of individual cytokines to the overall response is not always easy. One report indicates that the anti-proliferative action of IFNα against a human breast carcinoma cell line is actually mediated by TGFβ [28]; however, in some other systems TGFβ cannot mimic the effects of IFNα (M.J. Clemens, unpublished data). Cytokines also interact synergistically, particularly where one agent increases sensitivity to another by inducing its receptor (for example, IFNγ can induce the TNF receptor).

The tumor-inhibitory roles of cytokines *in vivo* include many regulatory effects on the immune system. For example, IFNs can induce expression of class I MHC antigens on many cell types, and this may allow more efficient recognition of tumor antigens by cytotoxic T cells. TNFα also acts against tumors by immune-specific mechanisms, as well as by causing impaired vascularization and consequent necrosis of the tissue – the property that gave this cytokine its name.

5.5.3 Cachexia

Many cancers, as well as chronic inflammation, some parasitic infections and other illnesses, result in weight loss, malnutrition and negative nitrogen balance, often with anemia. This debilitating combination of responses to the disease is known as cachexia [29]. There is good evidence that cytokines, especially TNFα (also called cachectin) mediate many features of cachexia. TNFα inhibits lipoprotein lipase and other enzymes required for fat storage by adipose tissue, resulting in depletion of body lipid stores. In addition, TNFα causes anorexia in experimental animals, contributing further to loss of body weight. Some (but not all) cancer patients have elevated serum TNFα levels, probably because of elevated production either by the tumor cells themselves or by associated lymphocytes and macrophages which may produce TNF as part of their attack on the tumor. Similar elevated serum TNFα levels are seen in some chronic parasitic infections that lead to cachexia such as malaria, and in AIDS patients, who are prone to developing cachexia. Thus much evidence suggests a role for TNFα in cachexia in man. However, there may well be roles for other cytokines, and IL-1 has been implicated in some of the mechanisms that lead to loss of body weight.

Another change associated with chronic inflammatory diseases (e.g. rheumatoid diseases), as well as the later stages of cancer, is degradation of bone, cartilage and connective tissue. Although cytotoxic cells contribute to these effects, some direct actions of cytokines are probably also involved. TNF and IL-1 induce connective tissue cells to produce collagenase and other proteolytic enzymes and inhibit production of new connective tissue components. The bone-resorbing activity of osteoclasts can be activated by IL-1β and the TNFs, and TGFα produced by tumors may also cause osteolysis and elevated calcium levels in the advanced stages of cancer.

5.6 Control of inflammation

In addition to their longer-term effects on cell growth and differentiation, on cell-mediated host defense mechanisms, and on chronic diseases like cancer, cytokines also mediate a number of important acute effects. Inflammatory responses are the best example of such short-term actions. Inflammation is a consequence of tissue damage and/or invasion by pathogens and is a complex reaction, not yet fully understood, involving many different cells and mediators [30]. It has three main features – pain, swelling and reddening – and the biochemical basis of these is discussed below.

Inflammation first involves vascular endothelial cells, macrophages, neutrophils and mast cells. T lymphocytes join in later when foreign antigens are presented to initiate immune responses. Attracting infiltrating cells (monocytes/macrophages and neutrophils) from the bloodstream and causing them to adhere to blood vessel walls and sites of tissue damage must require highly specific cell–cell interactions. These effects are achieved partly by release at the site of injury of chemoattractant agents such as IL-8 (see Chapter 1) and monocyte chemoattractant protein-1 (MCP-1), and partly by induction of adhesion molecules on the surfaces of neutrophils and endothelial cells.

These effects depend on several cytokines, including IL-1, the TNFs and (to a more limited extent) IFNγ. IL-1 and TNF also act on cells of the vascular endothelium to induce both procoagulant activity and increases in vascular permeability.

IFNs and TNFs also play a part in enhancing expression of MHC antigens by macrophages (and perhaps other cells too), permitting presentation of foreign antigens to T lymphocytes to initiate immune responses. Expression of class II MHC antigens is dependent on IFNγ. The subsequent production of a further wave of cytokines by activated T cells amplifies the original inflammatory responses, as well as leading to delayed hypersensitivity reactions at the site of injury.

When cytokines are released in sufficient quantities to enter the bloodstream, tissue inflammation can affect more distant organs *(Figure 5.7)*. CSFs can be produced and can stimulate bone marrow hematopoiesis, increasing the supply of appropriate blood cells to deal with the consequences of the injury. High levels of IL-1, TNFs or IFNs can act in the brain on the anterior hypothalamus to induce fever, and IL-6 production at the site of injury by macrophages, endothelial cells and other cell types signals the liver to synthesize and secrete the acute-phase proteins. The latter are also largely concerned with wound repair and protective functions.

5.6.1 Effects of prostaglandins and other inflammatory mediators

Many of the acute effects of inflammation are brought about by synthesis and release of small molecules – the inflammatory mediators [31]. Endothelial cells respond to

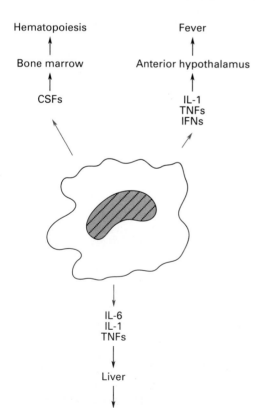

Figure 5.7: Systemic effects of cytokines in inflammatory diseases. Macrophages, lymphocytes and other cells activated during acute inflammation produce cytokines that can affect the function of distant organs. Induction of fever, stimulation of hematopoiesis and induction of the acute-phase proteins from liver are three examples. All these responses assist in combatting the cause of the inflammation and aid recovery.

stimulation by TNF or IL-1 by producing prostacyclin (PGI$_2$), a vasodilator. If produced in sufficiently large amounts, PGI$_2$ can lower blood pressure and cause shock. The same cytokines also induce PAF, which may also contribute to hypotension and increased capillary permeability. These mechanisms are believed to be particularly important in bacterial sepsis, when large amounts of TNFs and IL-1 may be released into the bloodstream. Another arachidonic acid metabolite, prostaglandin E$_2$, is probably the mediator of the hypothalamic response producing cytokine-induced fever.

5.6.2 The role of cytokines in allergic responses

Another important feature of local inflammatory responses is the release of mediators such as histamine, PAF, prostaglandins and leukotrienes by the basophilic granulocytes known as mast cells. Degranulation of mast cells (export from the cell of the preformed contents of granules) allows a rapid response to a variety of pathogenic stimuli. Mast cells are coated with immunoglobulin E (IgE) receptors. When IgE (previously produced in abnormally large amounts by B lymphocytes), which is bound to the receptors, encounters its specific antigen, degranulation is triggered, leading to an allergic response *(Figure 5.8)*. It is not clear why allergic individuals produce such large amounts of IgE specific for antigens which are often quite harmless [32].

Several cytokines have roles that may be important in the development of allergic responses. IL-4 enhances production of IgE by B lymphocytes and probably induces an immunoglobulin class switch to synthesis of IgE. This is further augmented by IL-5. IL-3 and IL-4 are also involved in controlling the proliferation of mast cell precursors.

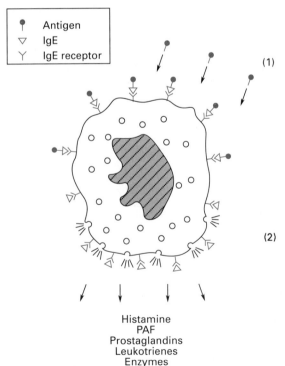

Figure 5.8: Mechanism of release of inflammatory mediators in allergic responses. Many of the symptoms of allergy are caused by the release of histamine and other inflammatory mediators by mast cells. When IgE bound to mast cell receptors encounters its specific antigen (1) the cells degranulate (2), releasing stored mediators.

The mast cells themselves produce IL-3 and IL-4, as well as other cytokines, when activated via the IgE receptor, providing a potential mechanism for autocrine cell proliferation in response to an allergic stimulus. Over-production of mast cells by this mechanism may, however, be limited by IFNγ since it inhibits production of IgE by B lymphocytes. The balance between IL-4 and IFNγ may determine the extent of an allergic response.

5.6.3 Chronic inflammation

Although inflammation is an essential host reaction to tissue injury, ensuring appropriate anti-microbial and tissue repair responses, over-stimulation of inflammatory pathways is pathogenic. Over-stimulation can be caused by unregulated production of cytokines at a site of injury, by abnormal cells such as some tumors, or by excessive release of inflammatory mediators in response to cytokines. The effects may include local damage or destruction of tissue and/or systemic actions such as induction of fever. Acute-phase protein synthesis by the liver may also be induced, although some acute-phase proteins may have a protective function by limiting the destruction of tissues by locally produced proteolytic enzymes.

In addition to these direct cytokine-mediated activities, macrophages present processed antigens to T lymphocytes, initiating immune responses which lead to generation of specific cytotoxic T cells, secretion of antibodies by B cells (see Chapter 4), and further production of cytokines. Again, unless these effects are limited they can lead to chronic inflammation or at least delayed hypersensitivity.

Chronic inflammation is often accompanied by wasting of the affected tissues. TNFα produced by macrophages may be involved in this, as in the cachexia associated with some cancers (see Section 5.5.3). IL-1 may also play a role. These two cytokines stimulate production of collagenase and other enzymes which destroy extracellular connective tissue, bone or cartilage. They also have the complementary effect of inhibiting the synthesis of new connective tissue matrix constituents, as well as inducing prostaglandin synthesis by synoviocytes. Thus the TNFs and IL-1 are believed to contribute to the damage to bones and joints characteristic of arthritis and other rheumatoid diseases [33]. Sometimes they can be detected at high levels in the synovial fluid from the joints of patients with rheumatoid arthritis. TNFα and IL-1 produced by alveolar macrophages may also cause some of the symptoms of pulmonary sarcoidosis, a granulomatous inflammatory lung disease that leads to fever and loss of body weight.

References

1. Sieff, C.A. (1990) *Ann. Rev. Med.*, **41**, 483.
2. Gough, N.M. and Williams, R.L. (1989) *Cancer Cells*, **1**, 77.
3. Axelrad, A.A. (1990) *Exp. Hematol.*, **18**, 143.
4. Dexter, T.M. and White, H. (1990) *Nature*, **344**, 380.
5. Rigby, W.F.C., Ball, E.D., Guyre, P.M. and Fanger, M.W. (1985) *Blood*, **65**, 858.
6. Lord, B.I. (1990) *Prog. Clin. Biol. Res.*, **352**, 391.
7. Mustoe, T.A., Pierce, G.F., Thomason, A., Gramates, P., Sporn, M.B. and Deuel, T.F. (1987) *Science*, **257**, 1333.
8. Foulis, A.K., Farquharson, M.A. and Meager, A. (1987) *Lancet*, **ii**, 1423.
9. Krown, S.E. (1990) *Biotherapy*, **2**, 137.
10. Xanthoudakis, S., Cohen, L. and Hiscott, J. (1989) *J. Biol. Chem.*, **264**, 1139.
11. Taylor, J.L. and Grossberg, S.E. (1990) *Virus Res.*, **15**, 1.

12. De Maeyer, E. and De Maeyer-Guignard, J. (1988) *Interferons and Other Regulatory Cytokines*. Wiley, New York.
13. Staeheli, P., Haller, O., Boll, W., Lindenmann, J. and Weissman, C. (1986) *Cell*, **44**, 147.
14. Rager-Zisman, B. and Bloom, B.R. (1985) *Br. Med. Bull.*, **41**, 22.
15. Sonenberg, N. (1990) *New Biol.*, **2**, 402.
16. Fauci, A.S. (1990) *Lymphokine Res.*, **9**, 527.
17. Bishop, J.M. (1991) *Cell*, **64**, 235.
18. Marshall, C.J. (1991) *Cell*, **64**, 313.
19. Pitha, P.M. (1990) *Cancer Cells*, **2**, 215.
20. Sawyers, C.L., Denny, C.T. and Witte, O.N. (1991) *Cell*, **64**, 337.
21. Depper, M., Leonard, W.J., Kronke, M., Waldmann, T.A. and Greene, W.C. (1984) *J. Immunol.*, **133**, 1691.
22. Sehgal, P.B. (1990) *Proc. Soc. Exp. Biol. Med.* **195**, 183.
23. Aaronson, S.A. and Pierce, J.H. (1990) *Cancer Cells*, **2**, 212.
24. Gullick, W.J. (1990) *Prog. Growth Factor Res.*, **2**, 1.
25. Besancon, F., Bourgeade, M.-F. and Testa, U. (1985) *J. Biol. Chem.*, **260**, 13074.
26. Taylor-Papadimitriou, J., Shearer, M. and Rozengurt, E. (1981) *J. Interferon Res.*, **1**, 401.
27. Tracey, K.J. and Cerami, A. (1990) *Ann. NY Acad. Sci.*, **587**, 325.
28. Kerr, D.J., Pragnell, I.B., Sproul, A. *et al.* (1989) *J. Mol. Endocrinol.*, **2**, 131.
29. Oliff, A. (1988) *Cell*, **54**, 141.
30. Arai, K., Lee, F., Miyajima, A., Miyatake, S., Arai, N. and Yokota, T. (1990) *Ann. Rev. Biochem.*, **59**, 783.
31. Hopkins, S.J. (1990) *Ann. Rheum. Dis.*, **49**, 207.
32. Herrod, H.G. (1989) *Ann. Allergy*, **63**, 269.
33. Verschure, P.J. and van Noorden, C.J. (1990) *Clin. Exp. Rheumatol.*, **8**, 303.

Further reading

General

Duff, G.W. (1989) Peptide regulatory factors in non-malignant disease. in *Peptide Regulatory Factors*. Edward Arnold, London, p. 112.

Anti-viral effects of the interferons

Langer, J.A. and Pestka, S. (1988) Interferon receptors. *Immunol. Today*, **9**, 393.

Cytokines and AIDS

Ensoli, B., Salahuddin, S.Z. and Gallo, R.C. (1989) AIDS-associated Kaposi's sarcoma: a molecular model for its pathogenesis. *Cancer Cells*, **1**, 93.

Cytokines and arthritic diseases

Feldmann, M., Brennan, F.M., Chantry, D. *et al.* (1990) Cytokine production in the rheumatoid joint: implications for treatment. *Ann. Rheum. Dis.*, **49**, 480.

Cytokines and inflammatory diseases

Revel, M. (1989) Host defense against infections and inflammations: role of the multifunctional IL-6/IFN-β2 cytokine. *Experientia*, **45**, 549.

Growth factors and growth inhibitors in cancer

Balkwill, F.R. (1989). *Cytokines in Cancer Therapy*. Oxford University Press, Oxford.

Steel, C.M. (1989) Peptide regulatory factors and malignancy. in *Peptide Regulatory Factors*. Edward Arnold, London, p. 121.

Hematopoiesis

Zipori, D. (1990) Regulation of hemopoiesis by cytokines that restrict options for growth and differentiation. *Cancer Cells*, **2**, 205.

Molecular basis of cancer

Lang, R.A. and Burgess, A.W. (1990) Autocrine growth factors and tumorigenic transformation. *Immunol. Today*, **11**, 244.

Origins of leukemia

Metcalf, D. (1989) The roles of stem cell self-renewal and autocrine growth factor production in the biology of myeloid leukemia. *Cancer Res.*, **49**, 2305.

Wound healing

Masure, S. and Opdenakker, G. (1989) Cytokine-mediated proteolysis in tissue remodelling. *Experientia*, **45**, 542.

6
THERAPEUTIC USES OF CYTOKINES

The previous chapter has indicated the many ways in which cytokines can be involved in regulating cellular activities essential to health. It is now clear that many diseases are due to disruptions of the normal pattern of cytokine activity. This raises the possibility of using cytokines clinically to intervene in the progress of diseases, or to alleviate symptoms or the side-effects of other therapies. With recombinant DNA technology, cytokine genes have been cloned and pure cytokines can be synthesized *in vitro* in large quantities. Nevertheless, this phase of cytokine biology is still in its infancy and relatively few uses for recombinant cytokines are yet in routine clinical practice.

Part of the problem is that, because so many cytokines have multiple actions *in vivo*, it is not always possible to predict the consequences of giving them to patients. Treatment with cytokines is seldom analogous to hormone replacement therapy (e.g. treatment of diabetes with insulin), because the relevant cytokine is rarely completely absent. Another imponderable is often the dose that should be given. High levels may not always be necessary (or even desirable) and, in the few clinical trials already performed, high doses of pure cytokines have sometimes produced unpleasant or dangerous side-effects. It should also be noted that recombinant human cytokines produced in bacterial, yeast or mammalian cell cultures may be structurally different from their normal counterparts; in particular they often lack the usual post-translational modifications (e.g. protein glycosylation) that may influence stability or tissue distribution *in vivo*.

Despite these problems the cytokines remain an important and potentially powerful class of agents (sometimes referred to as biological response modifiers) that will undoubtedly find a place alongside conventional drugs in the fight against diseases. This chapter will describe those uses already reported, and indicate areas where we may expect future applications.

6.1 Control of normal processes

One way in which cytokines may find widespread clinical application is for accelerating normal processes. This may be particularly useful when such processes are impaired as a result of other treatments that either affect endogenous cytokine production or alter the sensitivity of cells to the normally low levels of these factors in the body.

6.1.1 Wound healing

Wound healing is a complex process, involving mechanisms regulated by several growth factors and other cytokines *(Table 6.1)* (see Chapter 5). Since the activity of fibroblasts is central to wound repair, it is likely that any cytokine acting on fibroblasts could potentially accelerate (or, rarely, inhibit) healing. Thus the mitogenic growth factors PDGF, EGF and TGFα would be expected to promote wound repair. In the case of skin lesions, keratinocytes are known to be responsive to TGFα [1].

TGFβ may have a role as a mediator of wound repair; although it is a growth inhibitor for most cell types it has other effects which are beneficial. Indeed, even its anti-mitogenic actions may be important since they may provide a check to the hyperplastic effects of the growth factors mentioned above. It is particularly significant that TGFβ promotes laying down of connective tissue components such as collagen [2]. Increased collagen synthesis and the subsequent cross-linking of collagen chains will increase wound strength, a critical feature of the healing process. TGFβ also stimulates fibroblasts and monocytes to migrate to the site of a wound. These effects will also lead to other actions which help the damaged tissue to recover. In addition, TGFβ strongly promotes angiogenesis in the newly formed tissue. Animal studies support the notion that TGFβ therapy could accelerate wound repair [3], and this cytokine is the prime candidate for clinical application in this connection.

Table 6.1: Cytokines involved in wound healing

Cytokine	Actions in wound healing
PDGF	Proliferation of fibroblasts
EGF/TGFα	Proliferation of keratinocytes
TGFβ	Migration of fibroblasts and monocytes; restriction of local cell proliferation; stimulation of connective tissue synthesis; promotion of angiogenesis

6.1.2 Stimulation of hematopoiesis

In several clinical situations it would be advantageous to boost production of mature blood cells of various lineages *(Table 6.2)*. The principal applications that have received attention so far are the use of lineage-specific CSFs as replacement therapy in diseases where normal production of the factor is defective (especially the use of EPO to promote red cell production) [4], and treatment of cancer patients with CSFs to counter the myelosuppressive effects of chemotherapy [5].

Clinical trials have already shown that recombinant human erythropoietin (rhEPO) can increase the red cell mass in both normal individuals and patients who are EPO-deficient. The latter include those with severe renal disease and people with rheumatoid arthritis. Treatment with rhEPO reduces the severity of red cell anemia and diminishes dependence on blood transfusions. The factor is not antigenic, although it may have some side-effects such as hypertension and increased blood clotting. These may be results of the increased red cell mass relative to plasma volume. There is no apparent effect on kidney function in patients with renal failure, but a general improvement in well-being. The efficacy of EPO in these patients indicates its value as straight replacement therapy; future uses will probably include improving red cell production in autologous blood donors, premature babies, some patients with chronic infections or inflammation, and those who are severely anemic for other reasons.

CSFs have been used where bone marrow function is suppressed by chronic infection or after treatment with cytotoxic drugs. A group of AIDS patients treated with rhGM-CSF showed increases in blood counts of neutrophils, eosinophils and monocytes, although there was no clinical improvement [6]. Recombinant GM-CSF has also been shown to improve the rate of granulocyte recovery after autologous bone marrow transplantation in cancer patients who had previously been treated with high-dose chemotherapy [7]. Toxic effects are mostly limited to the consequences of monocyte/macrophage stimulation. Treatment of other cancer patients with rhG-CSF may have less toxic consequences and can reduce the neutropenia caused by chemotherapy. Such an effect may mean that higher doses of cytotoxic drugs, which are more effective against tumors, could be given to patients who receive rhG-CSF before or after treatment. Furthermore, improved levels of neutrophils should reduce the risk of serious infections, which are a common cause of death of cancer patients. Other patients who can benefit from GM-CSF or G-CSF are those with aplastic anemia or other disorders of bone marrow cell proliferation (e.g. myelodysplastic syndrome, a disease characterized by abnormally high numbers of poorly differentiated pre-leukemic blast cells). It may be advisable to test the cells for cytokine receptors that might allow the CSF to act as a growth factor, to avoid the risk that the treatment might make progression to acute leukemia more likely. For the same reason, CSF or IL-3 treatment to improve bone marrow function in leukemia patients may be unwise, since malignant cells derived from bone marrow lineages may accelerate their proliferation in response to these cytokines. However, some clinical studies suggest that this problem may not be universal [8].

Table 6.2: Uses of cytokines for stimulation of hematopoiesis

Cytokine	Clinical uses
EPO	Treatment of anemia caused by chronic renal failure; improvement of red cell counts following blood transfusions
GM-CSF	Improvement of granulocyte counts following bone marrow transplants; treatment of aplastic anemia; induction of terminal differentiation in myelodysplasia
G-CSF	Reduction of neutropenia following chemotherapy of cancer patients; similar uses to those for GM-CSF

A complementary approach might be to give lineage-specific growth-inhibitory factors such as the recently identified macrophage inflammatory protein-1α in conjunction with cancer chemotherapy. Such agents could temporarily block proliferation of hematopoietic stem cells; thus cytotoxic drugs specific for cells actively synthesizing DNA could attack tumor cells without affecting the normal bone marrow population.

Hematopoietic growth factors will no doubt become increasingly important in a variety of clinical applications. Possibilities include combined use of different CSFs and EPO, matched to the particular defects of the individual patient's bone marrow and timed to minimize possible cytotoxic effects of other therapies on normal progenitor cells. These new regimes will undoubtedly prove useful in reducing the side-effects of drug treatment, and in helping patients to resist opportunistic infections.

6.2 Regulation of immune function

Use of cytokines to modulate the immune system is becoming important in clinical medicine. Since many diseases are now recognized as disorders of immune regulation, involving depressed or inappropriate immune responses, it is logical to treat them with cytokines that stimulate or suppress lymphocyte functions. So far, however, this has been confined to cases where an entire class of cells is affected; we are not yet in a position to regulate the specificity of immune responses with cytokines.

6.2.1 Treatment of immunodeficiency syndromes

Attempts have been made to restore some of the depleted cytokines in patients suffering from immunodeficiency diseases. In AIDS the lack of functional T lymphocytes causes deficiencies of IL-2 and other interleukins, of IFNγ and of several CSFs, together with impaired production of IFNα in response to viral stimulation. Although IL-2 is an essential growth factor for T cells, it fails to restore T-cell numbers in AIDS patients and, by activating remaining infected T cells, may actually promote replication of HIV. In contrast, IFNα can be used successfully to treat Kaposi's sarcoma in some AIDS patients [9] and probably provides some protection against infection in this very vulnerable group. IFN is of little help in preventing replication of HIV or in restoring T-cell numbers *in vivo*, although it has an anti-viral effect against HIV *in vitro* and can act synergistically with azidothymidine [10].

IL-4 is an important growth factor for B and T cells and probably other cell types as well, so it might increase production of these cells if given exogenously. Unfortunately, however, it stimulates IgE production and may therefore cause allergic responses, making it unsuitable for treating immune disorders. Likewise, IL-1 and TNFα, which enhance various immune functions, probably have too many other effects and may prove too toxic for clinical applications. There may indeed be situations where it is desirable to supress TNFα. Together with IFNγ and IL-6, TNFα may be involved in rejection of transplanted organs. Additionally, acute graft-versus-host disease (where donor-derived T lymphocytes attack tissues such as gut and skin in an immunosuppressed host) may be caused by TNFα-mediated activation of cytotoxic cells. Use of antibodies against TNFα may therefore eliminate the disease, as suggested by data from some animal studies [11].

6.2.2 Activation of killer cells

The large granular lymphocytes called natural killer cells and LAK cells are under the control of cytokines *in vivo*. These cells are cytotoxic against virus-infected cells or tumor cells independently of MHC antigens. There has been much interest in the possibility that their ability to attack a tumor *in vivo* may be stimulated by the appropriate cytokine.

Natural killer cells can be activated by IFNα and β or IL-2 *in vitro* to exert cytotoxic activity against various target cells. The activated natural killer cells produce other cytokines such as TNFα, IFNγ and an agent known as natural killer cytotoxic factor, some of which may exert positive feedback on cells that secrete them. Stimulation of natural killer activity by administration of IFNs has no clear anti-tumor effect in cancer patients, perhaps because the cells are only transiently activated.

In contrast, activation of LAK cells by IL-2 has considerable clinical potential for treatment of tumors such as non-Hodgkin's lymphoma, melanoma and renal carcinoma [12]. IL-2 also has other immune-enhancing properties. It has been used alone or in combination with other cytokines (IFNs, TNF) or other forms of therapy. However, its use *in vivo* has been limited by serious side-effects including fever, nausea, tissue edema and hypotension. Many of these probably result from vascular leakage caused by adherence of activated lymphocytes to the vascular endothelium. There is also temporary liver dysfunction and marked myelosuppression (sometimes requiring transfusion to restore red cell levels). To avoid these problems, patients' lymphocytes are isolated by leukapheresis, treated with IL-2 *in vitro* and infused back into the circulation after activation. When combined with *in-vivo* IL-2 treatment, this approach may cause regression of tumors such as carcinomas of the colon or kidney, or malignant melanoma, that do not respond to other treatments [13]. Complete tumor remission is rare, and again side-effects can be severe (and occasionally fatal), due to the systemically administered IL-2 and perhaps also because of other cytokines produced by the activated LAK cells. A similar approach of treating cells with an activating lymphokine *in vitro* is being used to enhance the number and activity of cytotoxic tumor-infiltrating lymphocytes, obtained from a solid tumor, before infusion back into the patient. These new strategies will no doubt be refined and developed further, and potential hazards, such as accidental introduction of bacteria or viruses (e.g. hepatitis A) from the *in-vitro* cultures, minimized.

6.2.3 Control of autoimmune diseases

Little progress has yet been made in cytokine therapy of autoimmune diseases. This is largely because we lack knowledge of the specific roles cytokines play in causing these syndromes or in regulating progress of the disease. Excessive or abnormal T helper cell activity, or a relative deficiency of T suppressor cells, may underlie many of these conditions, therefore agents that control T lymphocyte functions may have clinical value. In animals, treatment with antibodies against the IL-2 receptor α sub-unit protected against development of autoimmune insulitis and renal damage [14]. Presumably this was a consequence of inhibiting T-cell activation by IL-2. The difficulty here will be to develop specific cell targetting, so that only T cells recognizing self-antigens are suppressed while normal protective immune responses are preserved.

In some multiple sclerosis patients intrathecal injections of IFNβ may reduce the frequency of disease exacerbations, but the mechanism is not clear. Even more surprising is the finding that IFNγ induces remissions in rheumatoid arthritis, since this cytokine induces expression of class II MHC antigens that might be expected to lead to increased presentation of self-antigens in this disease. However, IFNγ can cause increased production of cortisol that may be beneficial.

6.3 Infectious diseases

Since several cytokines, in particular the IFNs, are involved in normal host defenses they should be applicable to combat serious infections which do not respond to other therapies *(Table 6.3)*. However, the many unpleasant side-effects of systemic cytokine therapy preclude their use for mass prophylaxis against common infections like colds

Table 6.3: Infectious diseases suitable for IFN therapy

Hepatitis B
Genital herpes infections
Condylomata acuminata
Papillomas and other warts
Opportunistic infections in AIDS

or influenza, which are serious hazards to only a few people. Indeed many of the symptoms of these diseases are probably caused by IFNs and other cytokines produced in response to the virus.

Topical cytokines can be used to treat localized infections. For example, genital herpes improves following treatment with IFNα ointment. Usually, however, intramuscular or intravenous IFNs have been used. Herpes zoster and herpes labialis infections respond to intramuscular IFNα or IFNβ, and a number of virus-associated benign tumors have been treated in a similar way [15]. Papillomaviruses are known to cause growths such as juvenile laryngeal papillomas, genital warts *(Condylomata acuminata)* and other types of warts. These viruses may also contribute to the development of cervical cancers. The benign lesions asssociated with papillomaviruses respond well to IFN treatment, presumably due to direct anti-viral effects of these cytokines [16].

A major application of IFN therapy is the treatment of hepatitis B infections. In postneonatal infections IFNα has been shown to cause reductions in levels of the viral 'e' antigen and viral DNA in the blood and to alleviate hepatic inflammation [17], reducing the risk of liver cirrhosis. However, the response of hepatitis B acquired perinatally or in immunodeficient patients is less satisfactory. The effectiveness of IFNα against hepatitis B may be enhanced by combination with treatments which stimulate an immune response against virus-infected cells (e.g. IFNγ or IL-2) or which act in a different manner against replication of the virus (e.g. the anti-viral drug acyclovir). One potential limitation to the effectiveness of long-term IFN therapy is the development of neutralizing anti-IFN antibodies. However, such antibodies, where they appear, do not always reduce the efficacy of the treatment. Hepatitis B virus appears able to inhibit endogenous production of IFNα, thus treatment of infected individuals with this cytokine may represent a true example of replacement therapy. Since the disease

Table 6.4: Human malignancies responsive to IFN therapy

AIDS-related Kaposi's sarcoma
Carcinoid tumor
Carcinoma of the bladder
Chronic myelogenous leukemia
Cutaneous T-cell lymphoma
Hairy cell leukemia
Non-Hodgkin's lymphoma (low grade)

responds poorly to conventional treatments, IFN therapy is rapidly becoming the method of choice for managing this potentially life-threatening condition.

In view of the effectiveness of the IFNs against many pathogenic viruses *in vitro*, including HIV, it is disappointing that IFN therapy does little to reduce HIV replication or prevent infection of further cells in AIDS patients. This may be a consequence of the general impairment of the immune system. However, IFNs can cause regression of Kaposi's sarcoma (see Section 6.4) and reduce opportunistic infections in AIDS patients.

6.4 Cytokines in cancer therapy

It is clear that disorders of growth control, including all cancers, arise from disruption of the signalling pathways which normally control cell growth and differentiation. Many of these pathways are regulated by cytokines, hence attempts have been made to use cytokines to correct or counteract the abnormal behavior of cancer cells. Furthermore, since immune surveillance is important in the normal suppression of transformed cell growth, stimulation of surveillance by exogenous cytokines might give clinical benefits on top of the direct anti-tumor effects. Recent years have therefore seen numerous clinical trials of cytokine treatment of cancer, with varying rates of success.

6.4.1 Direct effects of growth inhibitors

In addition to the use of IL-2 to activate LAK cells (see Section 6.2.2), various cytokines have been tested as direct inhibitors of tumor cell growth. Excitement in the early 1980s that IFNs might be generally useful against many forms of cancer was largely generated by the media and was not scientifically warranted. Nevertheless the IFNs do directly inhibit proliferation of certain human tumor cells *in vitro* and can also have very impressive anti-tumor actions in experimental animal models. Disappointingly, these effects have not translated to the clinic, except in a few rare situations. The malignancies that do respond to IFNα or β therapy to some extent are mainly hematological ones *(Table 6.4)*; the common solid tumors such as carcinomas of the breast, lung or colon have been disappointingly refractory. In some cases this may be because the concentrations of IFN required to exert a direct anti-growth effect are not clinically achievable in solid tumors, but there are many tumor cell types whose growth is not inhibited by IFN, even in tissue culture.

The most successful use of IFNs against malignancy has been seen with the rare hairy cell leukemia [18]. Up to 80% of patients show partial or complete remissions after IFN therapy, with clearance of the leukemic B cells from the bone marrow in the latter group. This allows normal white cell counts to recover. It has been suggested that IFN acts against hairy cells by stimulating them to differentiate [19]; interestingly, IFNα also induces differentiation in other malignant human B-cell types such as chronic lymphocytic leukemias and the widely used Daudi Burkitt's lymphoma experimental cell line (see Chapter 5). The IFN doses given to hairy cell leukemia patients are limited by the side-effects (fever, malaise, myelosuppression and other symptoms) and possible development of anti-IFN antibodies. There is also some risk that rhIFNα may cause auto-immune disease. However, quite low doses of IFN, with minimal side-effects, are often effective against hairy cell leukemia, and the consequences of an immune response to one form of IFN may be overcome by changing to a mixture of naturally produced IFN species.

Chronic myelogenous leukemia (CML) also responds well to IFN treatment. IFN given during the chronic phase of the disease can achieve up to a 70% frequency of complete remission. Some non-aggressive lymphomas, such as the T-cell derived mycosis fungoides, are also treatable with IFNα, but higher-grade lymphomas respond poorly.

Of the solid tumors that have been examined for response to IFN therapy only carcinoma of the bladder, renal cell carcinoma, malignant melanoma, multiple myeloma and AIDS-associated Kaposi's sarcoma have shown partial remissions, and often in only a small percentage of cases. For some of these tumor types, more effective conventional therapies exist, although the results with melanoma are encouraging as this cancer is notoriously difficult to treat. The response to IFNα in patients with Kaposi's sarcoma is likely to be better if their immune functions are still largely intact.

In contrast to the partial successes with α or β IFNs, the clinical experience of using IFNγ in isolation has been disappointing. It does have some activity against CML, although it is not as effective as IFNα. However, IFNγ may have considerably greater potential when used in combination with IFNα or other cytokines with which it synergizes.

It is not clear what determines the relative sensitivity of different tumor cells to the anti-proliferative effects of IFNs. Most cells possess IFN receptors and show evidence of active signal transduction pathways (see Chapter 3). However, the consequences of IFN-induced signal transduction, in terms of the changes in gene expression that control cell growth, may well be different. Furthermore, the genetic changes that lead to the development of malignancy may negate the effects of new gene products induced by the IFNs *(Figure 6.1)*.

Other cytokines are being examined for anti-cancer activity in several clinical trials. Systemic treatment with TNFα has shown minimal effects, although intra-tumoral administration might be more effective where possible [20]. In view of the proposed mechanism of action of TNF (see Chapter 1) this may make sense; it has the added advantage of reducing some of the more unpleasant side-effects of high systemic doses of TNFα, which mimic the features of endotoxic shock. The combination of TNFα

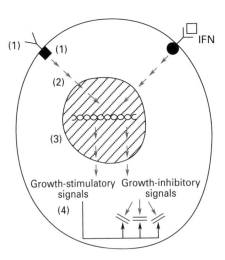

Figure 6.1: Antagonism between growth-promoting signals and the growth-inhibitory effects of IFNs. In normal cells, IFN induces changes in gene expression that may lead to growth inhibition. Tumor cells often have genetic dysregulation of growth factor receptor activity (1), aberrant signal transduction (2), inappropriate expression of growth-stimulatory genes (3), or over-activity of proteins that promote cell proliferation (4), all of which can make cell growth refractory to IFN treatment.

with IFNγ is also receiving attention. So far, little has been reported on possible anti-cancer uses of the other growth-inhibitory cytokine family, TGFβ and its relatives. However, the widespread anti-proliferative effect of TGFβ on cells in culture suggests potential therapeutic applications against tumors, as well as in wound healing (see Section 6.1). Again this might be more selective if the cytokine could be administered locally.

6.4.2 Combination with chemotherapy

There are promising indications that cytokines can be combined with chemotherapy to produce synergistic effects on tumor growth. These observations have several implications. First, treatment regimes may be identified in which a cytokine which is ineffective by itself enhances response to a drug. Secondly, for those drugs that are seriously toxic at high doses, the ratio of beneficial to harmful effects can be increased at a lower dose that is well tolerated. Thirdly, combinations of cytokines and cytotoxic drugs may be devised that exploit our knowledge of how each agent acts. Frequently, drugs used in cancer chemotherapy inhibit enzymes involved in DNA synthesis, whereas growth-inhibitory cytokines work via changes in signalling pathways and gene expression. Thus once we understand the biochemistry it may be possible to predict combinations of agents that interact synergistically. For example, recent clinical and experimental data show that the combination of IFNα or γ and 5-flourouracil is highly synergistic against some colorectal carcinomas [21, 22]. Knowing the biochemistry, we can suggest why this synergism happens *(Figure 6.2)*. In some cell types, IFN also inhibits uptake and phosphorylation of thymidine to give dTMP [23], perhaps allowing 5-fluorodeoxyuridylate to be a more effective enzyme inhibitor.

Several cytotoxic drugs have now been tested in combination with IFNs *(Table 6.5)*. Some combinations have been disappointing clinically despite favorable results in tissue culture or animal studies, and there is also the risk that IFN may adversely affect the response to some drugs (for example by accelerating their metabolism to inactive products). On the other hand, for this very reason IFN may also be able to protect normal tissues against the toxic effects of some compounds, allowing higher doses to

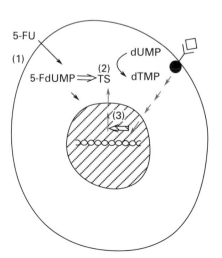

Figure 6.2: A putative model for the synergism between IFNs and 5-fluorouracil (5-FU) in inhibition of cell growth. 5-FU is taken up by cells and metabolized to 5-fluorodeoxyuridylate (1). 5-FdUMP inhibits thymidylate synthetase (TS) which converts uridylate (dUMP) to thymidylate (dTMP) (2), thus inhibiting DNA synthesis and cell proliferation. Cells can circumvent this by increasing synthesis of TS (3), but this response is blocked by IFN (hatched large arrow). Thus IFNs may sensitize cells to the anti-proliferative effects of 5-FU.

Table 6.5: Combinations of α IFNs and cytotoxic drugs in cancer therapy

Drug	Tumor treated
Chlorambucil	Low grade lymphoma
Cisplatin	Non-small cell carcinoma of the lung
Cyclophosphamide	Non-Hodgkin's lymphoma
Difluoromethylornithine	Melanoma
Doxorubicin	Several types of carcinoma
5-fluorouracil	Colorectal carcinoma
Melphalan/prednisone	Myeloma
Vinblastine	Renal cell carcinoma

be tolerated by the patient. Such effects, albeit mediated by different mechanisms, would be analogous to the protective effects of CSFs for bone marrow during chemotherapy (see Section 6.1).

6.4.3 Receptor antagonists

An alternative to using growth-inhibitory cytokines in cancer therapy is to interfere with the actions of positive growth factors, particularly where a tumor is factor-dependent for growth. Two strategies are possible, to attack the growth factor with neutralizing antibodies, or to attack the receptor or antagonize the interaction of the growth factor with its receptor *(Table 6.6)*. A disadvantage of the antibody approach is that it will not be specific for the tumor, but will interfere with all actions of the growth factor. Moreover, growth factors turn over rapidly *in vivo* and high levels of antibody would probably be needed to ensure continued inactivity of a factor. Attention has therefore been focused on targetting growth factor receptors. This might be more selective if the tumor possesses unusually high numbers of a particular receptor, and/ or if the tumor cells are more dependent on the growth factor for proliferation than are normal cells.

The EGF receptor is a suitable target since it may be over-expressed in various human tumors, especially gliomas. Anti-EGF receptor antibodies have been raised and conjugated to toxic proteins like ricin or to radioisotopes (for tumor imaging studies) [24]. Results indicated preferential uptake of these conjugates by tumors and inhibition

Table 6.6: Potential strategies for use of receptor antagonists

Neutralization of cytokine–receptor interaction with antibodies against either component
Use of soluble receptors to compete with cell surface receptors for cytokine binding
Design of inactive or toxic forms of cytokines that compete for receptor binding
Use of inhibitors of receptor signalling
Design of anti-receptor antibody–toxin conjugates
Creation of specific antibodies against mutant receptor forms

of EGF binding to the cells. In cases where a tumor expresses a mutated receptor (as in some gliomas), antibodies specific to the mutant receptor have been created, so that toxins or other therapeutic molecules are targetted at the tumor alone [24].

As an alternative to using an antibody, toxins can be conjugated to the growth factor itself, again with the aim that tumor cells expressing high levels of the EGF receptor will take up more of the toxic conjugate than normal tissues. This has been demonstrated for a human tumor xenograft grown in mice [25].

These new approaches will no doubt be extended to other growth factor receptors in due course, providing yet another way in which cytokine biology is exploited in the battle against cancer.

6.5 Inflammation

Some inflammatory diseases might be controlled by limiting the effect of overactive cytokine production, or providing an agent which can restore normal function. For example, chronic granulomatous disease may be treatable by administration of IFNγ. In this disease the respiratory burst oxidase system is defective in macrophages and neutrophils, making these cells unable to generate the reactive oxygen intermediates which kill infecting bacteria. The result is chronic inflammation at the sites of infection. IFNγ can restore the respiratory burst activity, allowing increased bacterial killing.

Allergy is another defect in immune regulation which might be correctable by IFNγ. Excessive production of IgE can be inhibited by this cytokine, with a concomitant switch towards IgG2a synthesis [26]. Thus the antagonistic effect of IFNγ towards the actions of IL-4 may be beneficial in the treatment of this inflammatory disorder.

As in cancer therapy, targetting of cytokine receptors may be a powerful strategy for controlling inflammation. For example, recent studies have identified naturally occurring antagonists of IL-1 that appear to act by competing with IL-1 for its receptors [27]. The gene for one of these has been cloned and found to encode a 152-amino acid protein that has the same affinity as IL-1α and β for the high affinity IL-1 receptor on fibroblasts. This antagonist, termed IL-1ra or IRAP by different groups, is synthesized by monocytes, is structurally related to the IL-1 family, and inhibits the biological effects resulting from interaction of IL-1 with its receptor on fibroblasts (e.g. PGE_2 production, hypotension and the acute-phase response) [28]. It also interferes with the adhesion of neutrophils and eosinophils to endothelial cells. In contrast, IL-1ra/IRAP does not bind to the low affinity IL-1 receptor of pre-B-cells, granulocytes and macrophages, and does not inhibit IL-1-induced neutrophilia *in vivo*. These findings suggest that it should be possible to inhibit different effects of IL-1 selectively using IL-1ra/IRAP. This could allow recombinant IL-1ra/IRAP to be used clinically to control inflammatory responses mediated by IL-1.

6.6 Future prospects for clinical applications of cytokines

Although the IFNs have been used clinically, at least in trials, for several years, we are still only at the beginning of the age of cytokine therapy for serious diseases. It is probably not an exaggeration to predict that this area is where we can expect the greatest impact of molecular biology and recombinant DNA on medical therapy in the immediate future. One can speculate on many ways in which growth factors and other cytokines, as well as growth factor antagonists, receptors and receptor antagonists will be developed and engineered in the coming years. Some of these ideas are summarized

Table 6.7: Possible developments in the use of cytokines and related molecules in clinical medicine

---

Development of mutant or chimeric cytokines with specific properties (competition with endogenous cytokines; presentation as toxic or radioactive conjugates; linkage to cell-specific antibodies)

Design of cytokine antagonists, based on information concerning the three-dimensional structure of the cytokine or its receptor

Synthesis of soluble forms of receptors, perhaps linked to antibodies or other ligands that localize the molecules to specific tissues.

Design of receptor antagonists utilizing protein structural information as above

Development of specific inhibitors of signal transduction pathways linked to engineered cytokine molecules that ensure delivery to the appropriate cell type

---

in *Table 6.7*. For instance, it is likely that further specificity will be introduced into cytokine structure, so that only certain cells bind the altered molecules; when conjugated with toxins, drugs or radioactive compounds these ligands may then be used to damage or destroy particular target cells. Alternatively, inhibitors of specific signal transduction pathways could be developed and delivered to the target cell by conjugation to the cytokine that normally activates that pathway.

It seems probable that the receptors will be exploited for novel forms of therapy, perhaps by designing soluble forms that compete for ligand binding or, more subtly, by synthesis of competitive inhibitors of cytokine–receptor binding. The latter approach will need much structural information about the shape of the cytokine and its site of interaction with the cell.

All of these approaches are also open to development by combining them with conventional chemotherapeutic treatments, and with the use of other cytokines or their antagonists. Thus we can anticipate an almost limitless range of highly sophisticated applications, based on our knowledge about the modes of action of the many agents that influence cell behavior in different situations. The success of these approaches is more likely to be restricted by society's willingness to pay for the fundamental work required to develop such therapies than it is by any limitation in the science that underpins this area of medicine.

References

1. Derynck, R. (1988) *Cell*, **54**, 593.
2. Roberts, A.B., Flanders, K.C., Kondaiah, P. *et al.* (1988) *Recent Prog. Hormone Res.*, **44**, 157.
3. Mustoe, T.A., Pierce, G.F., Thomason, A., Gramates, P., Sporn, M.B. and Deuel, T.F. (1987) *Science*, **257**, 1333.
4. Dessypris, E.N., Graber, S.E., Krantz, S.B. and Stone, W.J. (1988) *Blood*, **72**, 2060.
5. Balkwill, F.R. (1989) *Cytokines in Cancer Therapy*. Oxford University Press, Oxford.
6. Groopman, J.I., Ronald, M.D., Deleo, M.J. and Golde, D.W. (1987) *New Engl. J. Med.*, **317**, 593.
7. Devereux, and D. Linch, D.C. (1989) *Br. J. Cancer*, **59**, 2.
8. Vadhan-Raj, S., Keating, M., LeMaistre, A. *et al.* (1987) *New Engl. J. Med.*, **317**, 1545.
9. Mitsuyasu, R.T. (1989) *Interferons Cytokines*, **12**, 6.

10. Dubreuil, M., Sportza, L., D'Addario, M., Lacoste, J., Rooke, R., Wainberg, M.A. and Hiscott, J. (1990) *Virology*, **179**, 388.
11. Piguet, P.F., Grau, G.E., Allet, B. and Vassalli, P. (1987) *J. Exp. Med.*, **166**, 1280.
12. Hancock, B.W. and Rees, R.C. (1990) *Cancer Cells*, **2**, 29.
13. Semenzato, G. (1990) *Leukemia*, **4**, 71.
14. Kelley, V.E., Gaulton, G.N., Hattori, M., Ikegami, H., Eisenbarth, G. and Strom, T.B. (1988) *J. Immunol.*, **140**, 59.
15. Taylor-Papadimitriou, J. (1985) *Interferons: Their Impact in Biology and Medicine*. Oxford University Press, Oxford.
16. Healy, G.B., Gelber, R.D., Trowbridge, A.L. *et al.* (1988) *New Engl. J. Med.*, **319**, 401.
17. Jacyna, M.R. and Thomas, H.C. (1990) *Br. Med. Bull.*, **46**, 368.
18. Thompson, J.A. and Fefer, A. (1987) *Cancer*, **59**, 605.
19. Michalevicz, R. and Revel, M. (1987) *Proc. Natl Acad. Sci. USA*, **84**, 2307.
20. Taguchi, T. (1987) *Proc. ASCO*, **6**, 233.
21. Wadler, S. and Schwartz, E.L. (1990) *Cancer Res.*, **50**, 3473.
22. Chu, E., Zinn, S., Boarman, D. and Allegra, C.J. (1990) *Cancer Res.*, **50**, 5834.
23. Gewert, D.R., Shah, S. and Clemens, M.J. (1981) *Eur. J. Biochem.*, **116**, 487.
24. Harris, A.L. (1990) *Cancer Cells*, **2**, 321.
25. Heimbrook, D.C., Stirdivant, S.M., Ahern, N.L. *et al.* (1990) *Proc. Natl Acad. Sci. USA*, **87**, 4697.
26. Coffman, R.L. and Carty, J. (1986) *J. Immunol.*, **136**, 949.
27. Whicher, J. (1990) *Nature*, **344**, 584.
28. Postlethwaite, A.E., Raghow, R., Stricklin, G.P., Poppleton, H., Seyer, J.M. and Kang, A.H. (1988) *J.Cell Biol.*, **106**, 311.

Further reading

Control of hematopoiesis

Metcalf, D. (1989) Haemopoietic growth factors 2: clinical applications. in *Peptide Regulatory Factors*. Edward Arnold, London, p. 25.

Cytokines and wound healing

Hudson-Goodman, P., Girard, N. and Jones, M.B. (1990) Wound repair and the potential use of growth factors. *Heart Lung*, **19**, 379.

Cytokines in inflammatory diseases

Arend, W.P. and Dayer, J.M. (1990) Cytokines and cytokine inhibitors or antagonists in rheumatoid arthritis. *Arthritis Rheum.*, **33**, 305.

Interferons and cancer therapy

Liberati, A.M., Fizzotti, M., Di Clemente, F. *et al.* (1990) Response to intermediate and standard doses of IFN-b in hairy-cell leukaemia. *Leukobr. Res.*, **14**, 779.

Other cytokines in cancer treatment

Russell, S.J. (1990) Lymphokine gene therapy for cancer. *Immunol. Today*, **11**, 196.

Receptor antagonists

Rubin, L.A. and Nelson, D.L. (1990) The soluble interleukin-2 receptor: biology, function, and clinical application. *Ann. Intern. Med.*, **113**, 619.

Treatment of immunodeficiency diseases

Mitsuyasu, R.T., Miles, S.A. and Golde, D.W. (1990) The use of myeloid hematopoietic growth factors in patients with HIV infection. *Int. J. Cell Cloning*, **8**, 347.

Treatment of infectious diseases

Murray, H.W. (1990) Interferon-gamma therapy in AIDS for mononuclear phagocyte activation. *Biotherapy*, **2**, 149.

Use of natural killer and LAK cells

Richards, J.M. (1989) Therapeutic uses of interleukin-2 and lymphokine-activated killer (LAK) cells. *Blood Rev.*, **3**, 110.

APPENDIX A. GLOSSARY

Adhesion molecule: cell surface protein or glycoprotein required for intercellular adhesion; often inducible by cytokines or other extracellular influences.

Angiogenesis: development of a blood supply to new tissue; can be promoted by certain cytokines.

Antibody: immunoglobulin molecule synthesized in response to a specific antigen and which binds to that antigen. Several classes (isotypes) exist.

Antigen: any molecule that can stimulate the production of a specific antibody directed against it.

Autocrine stimulation: process by which a hormone or cytokine acts on the same cells that produce it.

Cachexia: state characterized by loss of body weight, negative nitrogen balance and wasting of many tissues.

Calmodulin: calcium-binding protein involved in regulation of some protein kinases.

Chemotaxis: process by which cells migrate to a new location in response to specific stimuli released by other cells.

Chemotherapy: treatment of a disease by means of chemical inhibitors of processes such as DNA synthesis.

Colony stimulating factors: cytokines that regulate one or more pathways of hematopoiesis in the bone marrow. Some colony stimulating factors also have other biological activities in other tissues.

Cyclic AMP: second messenger molecule synthesized by adenylate cyclase which regulates the activity of a family of protein kinases.

Cytokine: protein or glycoprotein produced by one or more cell types that regulates the activity of other cells.

Diacylglycerol: second messenger molecule produced by the hydrolysis of phospholipids; exerts its effect by activation of the protein kinase C family.

Eicosanoids: molecules derived from arachidonic acid that regulate a variety of functions associated with inflammatory responses.

Enzyme-linked immunosorbent assay: a sensitive means of quantifying an antigen by binding a specific antibody coupled to an enzyme that can synthesize an easily measurable product.

Erythropoiesis: process of proliferation and differentiation of the red blood cell lineage.

Exon: region of a gene that is transcribed and appears in the mature RNA product; includes the protein-coding sequences of messenger RNAs.

Glycoprotein: protein covalently modified by addition of one or more carbohydrate side-chains.

Glycosylation: process of covalent addition of carbohydrates to a protein or other acceptor.
Growth factor: naturally occurring agent that stimulates the proliferation and/or differentiation of cells.
Hematopoiesis: process of proliferation and differentiation of blood cell lineages.
Homeostasis: maintenance of constant conditions in cells and body fluids.
Inositol phosphates: phosphorylated forms of inositol, usually derived from the hydrolysis of inositol phospholipids, which act as second messengers for the control of cytoplasmic calcium concentrations.
Interferons: (glyco)proteins that are synthesized in response to virus infections or antigenic stimulation and which confer a state of resistance to virus infection on their target cells. The interferons also exert a variety of other actions.
Interleukins: proteins produced by a variety of cell types in response to antigenic stimulation and which regulate many functions of the immune system as well as other processes.
Intron: region of a gene that is transcribed but is spliced out in creating the mature RNA product.
Lymphokine-activated killer cell: a specialized type of lymphocyte that is activated by lymphokines such as IL-2 and is cytotoxic towards foreign cell types including tumor cells.
Lymphocyte: specialized class of cells that recognizes and discriminates between different antigens; classified into T and B lymphocytes on the basis of developmental and functional criteria. T cells recognize processed antigens and respond by producing a variety of cytokines; B cells produce and secrete specific immunoglobulins (antibodies).
Lymphokine: class of cytokine produced by cells of the immune system (especially T lymphocytes).
Macrophage: specialized cell which ingests foreign cells and particles by phagocytosis and processes their proteins to produce peptide antigens that are presented to T lymphocytes. Macrophages also produce a variety of cytokines.
Major histocompatibility complex: genes encoding a family of cell surface glycoproteins involved in presentation of antigens to other cell types; very important in the distinction between self and non-self by the immune system.
Mitogen: any agent that stimulates cell growth and division.
Monocyte: precursor of the macrophage.
Natural killer cell: specialized lymphocyte that kills virus-infected and tumor cells by a non-antibody-dependent process.
Oncogene: gene encoding a protein involved in cell growth control which, when mutated or over-expressed, can contribute to the transformation of cells into tumor cells.
Paracrine stimulation: process by which a hormone or cytokine acts on a different cell type which lies in close proximity to the producer cells.
Promoter: region of a gene that constitutes the start site for RNA synthesis.
Protein kinase: enzyme that phosphorylates protein substrates on serine, threonine or tyrosine residues.
Protein phosphorylation: post-translational modification of a protein involving the enzyme-catalyzed addition of phosphate groups to the hydroxyl side-chains of serine, threonine or tyrosine residues.
Proteolysis: cleavage of the polypeptide backbone of a protein to yield smaller peptides.

Proto-oncogene: normal cellular equivalent of an oncogene (q.v.), involved in the control of cell proliferation or differentiation.

Radioimmunoassay: a sensitive means of quantifying an antigen by binding a radioactively labeled specific antibody.

Receptor: protein(s) on the cell surface or within the cell that specifically binds and responds to a cytokine or hormone.

Scatchard plot: of bound/free versus bound ligand concentrations that permits calculation of the number of binding sites and the affinity of those sites for the ligand.

Signal peptide: sequence on a protein, usually at the N-terminus, that causes binding of the nascent protein and ribosome to membranes of the endoplasmic reticulum and brings about passage of the newly synthesized protein into the lumen of the endoplasmic reticulum; normally present on proteins destined for secretion and removed shortly after passage of the nascent chain through the ER membrane.

Signal transduction: initial process whereby the binding of a cytokine to its receptor results in changes in the biological activity of the cell.

Stem cell: precursor cell that proliferates and gives rise to more differentiated progeny (e.g. in the bone marrow).

Stromal cells: connective tissue cells in the bone marrow that interact with and regulate the activity of blood cell precursors.

Transcription factors: proteins that regulate RNA synthesis by interacting with specific regions of DNA in the nucleus.

Transforming growth factors: growth factors produced by both normal and tumor cells that can contribute to the transformed phenotype of the latter; historically classified into TGFα and TGFβ, although these factors are unrelated in structure or function.

Tumor necrosis factors: cytokines originally associated with the ability to cause cell death and necrosis in tumors but having many additional properties.

Tumor suppressor gene: a gene, the loss of expression or mutation of which can contribute to transformation of cells to a tumorigenic phenotype; believed to be important in the negative control of proliferation in normal cells.

INDEX

Activins, 15, 24
Acute-phase proteins, 9, 93, 95, 109
Acyclovir, 104
Adenylate cyclase, 42, 47
Adhesion, 109
Adhesion molecules, 51, 92
AIDS, 61, 81, 92, 101, 102, 105
Allergic
　disease, 9,109
　response, 8, 94, 102
Anemia
　aplastic, 101
　red cell, 100
Angiogenesis, 30, 61, 80, 89, 100
Anti-interferon antibodies, 104, 105
Anti-oncogenes, 63, 65
Antibody production, 15, 71
Antigen presentation, 68
Antigens, 26, 29
Arachidonic acid, 16, 44, 47, 48, 91, 94
Arthritis, 95
Aspirin, 47
Assays, 30
Autocrine regulation, 1, 7, 13, 15, 18, 27, 60, 65, 69, 71, 83, 89, 91
Autoimmune diseases, 80, 82, 103, 105
Azidothymidine, 102

B cell growth factors, 70
B lymphocytes, 61, 70, 71
Bacterial sepsis, 94
BFU-E, 13
Binding proteins, 31

Biological assays, 30
Biological response modifiers, 99
Bone, 14, 24, 92, 95
Bone morphogenetic proteins, 15, 24
Bone marrow, 9, 10, 13, 61, 66, 75, 101
Bone marrow transplantation, 101
Burkitt's lymphoma, 51, 52

Cachectin, 92
Cachexia, 86, 92
Calcium, 29, 43, 44, 69, 70
Calmodulin, 44, 49
Cancer, 75, 85, 92, 100, 101
Cancer therapy, 105, 108
Carcinoma
　breast, 89, 91
　colorectal, 107
　renal cell, 106
Cartilage, 15, 95
Cathepsin D, 28
CD23, 71
Celiac disease, 81
Cell cycle, 62, 63, 65, 66, 70, 90
Cell differentiation, 18, 51, 52, 62, 65, 66, 75
Cell killing, 16
Cell proliferation, 18, 62, 75
Cell surface proteins, 51
Cell transformation (*see* Transformation)
CFU-E, 13
CFU-GEMM, 77
CFU-GM, 77
Chemotaxis, 10, 12, 15, 80

Chemotherapy, 101, 107, 110
Chronic granulomatous disease, 109
Chronic inflammation, 95
Cloning, 62
Collagen, 80, 91, 100
Collagenase, 92, 95
Colony stimulating factors, 10–12, 61, 65–67, 77, 78, 93, 101, 102
Commitment, 63, 69
Common cold, 103
Connective tissue, 15, 80, 95, 100
CSF-1, 12
Cyclic AMP, 42
Cyclo-oxygenase, 47
Cytokine, 1
 genes, 21
 networks, 57
 precursors, 27–29
 production, 25,29
Cytotoxic drugs, 107
Cytotoxic T cells, 70, 72, 91,102
Cytotoxicity, 12, 18

Daudi cells, 65, 105
Delayed hypersensitivity, 93, 95
Degradation of mRNA, 29
Diabetes mellitus, 81
Diacylglycerol, 43, 46
Differentiation antigens, 52
Dissociation constants, 30
DNA replication, 62, 63, 69, 91, 107
Double-stranded RNA, 26, 82, 83
Down-regulation of receptors, 30, 60, 66, 67

EGF, 4, 14, 23, 43, 79, 100
EGF receptor, 87,108
Eicosanoids, 47
eIF-2, 83
ELISA, 31
Embryonic development, 14, 15
Endocytosis, 37
Endoplasmic reticulum, 44
Endothelial cells, 66, 80, 92, 93
Endotoxic shock, 106
Endotoxin, 12, 16
Enhancers, 21
Enzyme-linked immunoassays, 31

Eosinophils, 9
Epidermal growth factor (*see* EGF)
Epstein–Barr virus, 82, 84
ERBB, 87
Erythropoiesis, 78
Erythropoietin, 13, 77, 78, 100, 101
Estrogens, 13
Evolutionary relationships, 23
Exons, 21
Extracellular matrix, 91

Fc receptor, 18, 72
Febrile response, 9
Feedback control, 8
Fever, 47, 85, 93–95
FGR, 51, 65
Fibroblast growth factors, 4, 24, 61, 79, 87, 89
Fibroblasts, 66, 100
Fibronectin, 80
Fluorouracil, 107
FMS, 13, 67, 87
FOS, 49, 63, 65, 69, 70, 87
Free radicals, 16, 47, 91

G proteins, 41, 43, 86
G-CSF, 9, 24, 77, 101
Gene expression, 48
Genital herpes, 104
Genital warts, 104
Glucose transport, 90
Glycosylation, 26
GM-CSF, 8, 10, 23, 24, 61, 77, 82, 85, 88, 101
Graft-versus-host disease, 102
Granulocytes, 11
Growth factors, 3, 62, 63
Growth inhibition, 15, 65
Growth inhibitors, 89, 90, 101, 105

Hairy cell leukemia, 52, 66, 105
Hematopoiesis, 15, 27, 66, 75, 93, 100
Hematopoietic cells, 10, 16, 65
Hepatitis B, 104
Herpes
 labialis, 104
 viruses, 84
 zoster, 104

Histamine, 94
Histocompatibility antigens, 16, 18, 52, 80
HIV, 61, 81, 82, 84, 85, 102, 105
Homeostasis, 75
Hypotension, 109
Hypothalamus, 94

IFN (*see* Interferon)
IgE, 71, 94
IGF-binding proteins, 28
IGF-I, 23, 63, 89
IGF-II, 5, 23, 26
IL-1, 6, 24, 26, 30, 43, 49, 57, 61, 69, 71, 77, 78, 90, 91–93, 95, 10
IL-1 receptor, 6, 109
IL-2, 15, 24, 27, 30, 49, 65, 69, 70, 88, 102–104
IL-2 receptor, 7, 35, 38, 69, 71, 88, 103
IL-3, 8, 10, 23, 26, 67, 77, 78, 88, 94
IL-3 receptor, 8
IL-4, 8, 23, 24, 26, 30, 53, 70, 71, 77, 89, 94, 102
IL-4 receptor, 8, 89
IL-5, 9, 23, 24, 26, 77, 78, 94
IL-5 receptor, 9
IL-6, 9, 17, 24, 69, 71, 77, 85, 89, 93, 102
IL-6 receptor, 9
IL-7, 9
IL-8, 10, 92, 105
Immune system, 68, 75, 80, 91, 95, 102
Immunoassays, 31
Immunodeficiency, 102
Immunoglobulin isotype switching, 71
Immunoglobulin superfamily, 6, 9, 36
Indoleamine 2,3-dioxygenase, 91
Infection, 71, 75, 103
Inflammation, 12, 61, 71, 75, 109
Inflammatory
  agents, 47, 48, 93, 95
  conditions, 9, 92
  responses, 15, 16, 79, 92
Influenza, 84, 104
Inhibins, 15, 24
Inositol phosphates, 43, 44, 70
Insulin, 5, 23, 63
Insulin-like growth factors (*see* IGF-I and IGF-II)
Interferons, 17, 21, 29, 75, 78, 80, 82, 90, 93, 103–106, 109
INT2, 24, 87, 89
Interferon (acid labile), 80, 82
Interferon α, 24, 26, 65, 71, 81, 82, 87, 102, 107
Interferon β, 26, 65, 71, 87, 102, 103
Interferon γ, 8, 24, 52, 59, 71, 78, 81, 85, 91, 93, 95, 102–104, 106, 107, 109
Interferon-induced protein kinase, 83, 84
Interleukins, 6, 68
Interleukin-1 (*see* IL-1, etc.)
Introns, 18, 24
IRAP, 109

JUN, 87
Juvenile laryngeal papilloma, 104

Kaposi's sarcoma, 61, 82, 89, 102, 105, 106
Keratinocytes, 80, 100
Kidney, 13, 14

LAK cells, 91, 102
LCK, 70
Lectins, 69
Leukapheresis, 103
Leukemia
  acute, 88, 101
  chronic myelogenous, 106
Leukemia inhibitory factor, 14
Leukotrienes, 44, 48, 94
Lipopolysaccharide, 6, 9, 12, 15
Lipoprotein lipase, 92
Lipoxygenase, 48
LPS (*see* Lipopolysaccharide)
Lymphocyte activation, 38, 68, 102, 103
Lymphoma, 88, 89, 106
Lymphokines, 5
Lymphotoxin, 16

M-CSF, 23, 24, 77
M-CSF receptor, 67, 87
Macrophage inflammatory protein-1α, 77, 101
Macrophages, 26, 52, 66, 68, 71, 91, 92, 95, 109
Malaria, 92
Malignant melanoma, 106

Mast cells, 8, 92, 94
MCP-1, 92
Megakaryocytes, 13
Membrane-bound proteins, 27
MHC antigens
  class I, 71, 91
  class II, 69, 72, 93, 103
Mitosis, 62
Monoclonal antibodies, 31, 62
Monocytes, 66, 71, 109
Monokines, 6
Morphogenesis, 30
mRNA stability, 26, 51
Müllerian inhibitory substance, 15, 24
Multiple sclerosis, 81, 103
Mx protein, 84
MYB, 70
MYC, 51, 63, 65, 68–70, 91
Mycosis fungoides, 106
Myelodysplastic syndrome, 101
Myeloma, 89, 106

Natural killer cytotoxic factor, 102
Natural killer cells, 18, 84, 91, 102
Nerve growth factor, 5
Neutrophils, 10, 11, 92, 101, 109
NF-κB, 50, 85

Oligoadenylate synthetase, 83
Oncogenes, 13
Osteoclasts, 92
Osteolysis, 86

p53, 87
Paracrine regulation, 2, 11, 18, 27, 60, 78, 83, 89
PDGF, 4, 15, 23, 43, 61, 63, 79, 87, 90, 100
PDGF receptor, 23, 35
Peroxide, 13
Phagocytosis, 12, 13, 16
Phorbol esters, 9, 46
Phosphatidylcholine, 43, 46
Phosphatidylinositol 4,5-bisphosphate, 43
Phosphoinositide kinase, 46
Phospholipases, 43, 86
Phospholipids, 69
Phytohemagglutinin, 7

Placenta, 13, 14
Plasma cells, 71
Plasmin, 28, 80
Plasminogen activator inhibitor-1, 80
Platelet-activating factor, 48, 94
Platelet-derived growth factor (see PDGF)
Platelets, 13, 14, 80
Poliovirus, 84
Poly(A), 21
Post-translational modifications, 26, 99
Precursors, 27–29
Progression factors, 63, 71
Promoters, 21
Prophylaxis, 85
Prostacyclin, 44, 47, 94
Prostaglandins, 13, 30, 44, 47, 93–95, 109
Proteases, 28, 91
Protein kinase A, 49
Protein kinase C, 29, 43, 46, 49, 67, 69, 70
Protein phosphatases, 49, 86
Protein phosphorylation, 25, 26, 39, 42, 44, 49, 50, 63, 65, 69, 70, 86
Protein synthesis, 83, 91
Proteoglycans, 80
Proteolysis, 23
Proto-oncogenes, 24, 40, 63, 86
Pulmonary sarcoidosis, 95

Radioimmunoassay, 31
RAS genes, 41, 87
RB1 gene, 65, 66, 87
Receptor
  antagonists, 108, 110
  down-regulation, 30, 60, 66, 67
  internalization, 37
  transmodulation, 60
Receptors, 29, 33, 91, 109
Red cells, 100
Renal disease, 100
Respiratory burst activity, 109
Response elements, 21, 25, 26, 49
Retroviruses, 39, 63
Rheumatoid diseases, 95, 100, 103
Ricin, 108

Scatchard plot, 34
Second messengers, 38, 41, 49
Secretion, 26, 27

Shock, 47, 94
Side-effects, 99, 100, 102, 103, 105, 106
Signal sequence, 7, 26
Signal transduction, 25, 29, 38, 63, 65, 67, 69, 70, 105, 106, 110
SIS, 87, 90
Soluble receptors, 29, 110
Stem cells, 75
Stromal cells, 27, 61, 78
Superoxide, 12
Surface immunoglobulin, 70
Synergism, 67
Systemic lupus erythematosus, 80, 82

T cell antigen receptor, 26, 69
T helper cells, 81
T lymphocytes, 24, 26, 29, 30, 49, 65, 66, 71, 92
Tac antigen, 7, 71
Terminal differentiation, 68, 71, 89, 105
TGFα, 14, 23, 79, 89, 92, 100
TGFβ, 14, 23, 24, 28, 30, 57, 61, 65, 70, 71, 77, 79, 80, 90, 91, 100, 107
Therapeutic uses, 99
Thromboxanes, 44, 48
TNF (*see* Tumor necrosis factors)
Tissue
  damage, 71
  rejection, 102
  repair, 15
Tissue-specific inhibitor of metalloprotease, 80

Toxins, 109, 110
Trans-acting factors, 26, 49
Transcription, 21, 24, 29, 49, 50
Transcription factors, 63, 69, 87
Transferrin, 69
Transformation, 24, 86, 87, 89, 90
Transforming growth factors (see TGFα and TGFβ)
Translational control, 26, 50
Transplantation, 102
Tumor growth, 61, 107
Tumor-infiltrating lymphocytes, 103
Tumor necrosis factors, 15, 23, 30, 52, 60, 65, 70, 71, 78, 80, 85, 90–93, 95, 102, 106
Tumor promoters, 46
Tumor suppressor genes, 63, 66, 87, 90
Tumors, 14, 15, 103
Tyrosine kinases, 13, 39, 43, 49, 65, 67, 69, 86

Uterus, 13

v-*erbB*, 89
v-*sis*, 90
Vaccinia virus, 14
Vascular endothelial cells, 66, 80, 92
Viral infection, 18, 82
Viruses, 9, 26, 30, 82

Wasting, 95
Wound healing, 14, 15, 28, 75, 79, 99

Foreword

The **Information Systems Engineering Library** provides guidance on managing and carrying out Information Systems Engineering activities. In the IS lifecycle, Information Systems Engineering takes place once the IS strategy has been defined. It is concerned with the development and ongoing improvement of information systems up to the operational stage, and their maintenance whilst in operational use.

The Information Systems Engineering Library complements other CCTA products, in particular the project management method, PRINCE, and the systems analysis and design method, SSADM.

Volumes in the Information Systems Engineering Library are of interest to varying levels of staff from IS directors to IS providers, helping them to improve the quality and productivity of their IS development work. Some volumes in this library should also be of interest to business managers, IS users and those involved in market testing, whose business operations depend on having effective IS support by means of Information Systems Engineering activities.

The Information Systems Engineering Library also complements other related CCTA publications, particularly the Programme and Project Management Library, the Information Management Library for data management issues, the IT Infrastructure Library for operational issues and the IS Planning Subject Guides for strategic issues.

CCTA welcomes customer views on Information Systems Engineering Library publications. Please send your comments to:

>   Information Systems Engineering Group
>   CCTA
>   Rosebery Court
>   St Andrews Business Park
>   Norwich
>   NR7 0HS

Information Systems Engineering Library
Reuse in SSADM using Object-Orientation

Contents

Chapter		Page
1	**Introduction**	7
	1.1 Purpose of this volume	7
	1.2 Who should read this volume	8
	1.3 Status of this volume	8
	1.4 Associated publications	9
	1.5 Structure of this volume	9
2	**OO concepts**	11
	2.1 Introduction	11
	2.2 Classes and objects	11
	2.3 Encapsulation	12
	2.4 Class hierarchies and inheritance	14
	2.5 Message-passing	15
	2.6 Methods and polymorphism	15
	2.7 Benefits and application of OO concepts	15
	2.8 Applicability of OO concepts to the development of information-based systems	17
3	**OO concepts in SSADM**	19
	3.1 Introduction	19
	3.2 OO paradigms	19
	3.3 Role of the OO paradigms in SSADM	20
	3.4 Implementation independence of Conceptual Model	24
	3.5 Objects as 'long-running' processes	25
	3.6 Encapsulation in SSADM	25
	3.7 Composite objects and methods, and aggregate objects	27
	3.8 Class hierarchies in SSADM	30
4	**Data modelling extensions**	35
	4.1 Introduction	35
	4.2 Class hierarchies	37
	4.3 Aspects	43
	4.4 Discovering class hierarchies	47
	4.5 Development of the Logical Data Model	48
	4.6 Using aspects to prepare the LDM for detailed development	52
	4.7 Example of achieving reuse	59
	4.8 Effect of object behaviour on reuse	63
	4.9 Limits to Reuse	65
	4.10 Validating the LDM by entity-event modelling	68

5	**Process modelling extensions**		**71**
	5.1	Introduction	71
	5.2	Methods and operations	73
	5.3	Entity state optimisation	76
	5.4	Discovering the 'right' set of events or methods	78
	5.5	Identification and analysis of reusable methods	82
	5.6	Entity class hierarchies and inheritance	95
	5.7	Encapsulation of methods for an entity type	101
	5.8	Effect correspondences	102
	5.9	Enquiry modelling	106
6	**Method invocation and message-passing**		**109**
	6.1	Introduction	109
	6.2	Deriving message-passing requirements	109
	6.3	Two approaches for implementation	112
	6.4	Implications of using network invocation	113
	6.5	Implications of using direct invocation	117
	6.6	Summary of message-passing approaches	121
7	**External and Internal Design**		**125**
	7.1	Introduction	125
	7.2	External Design	125
	7.3	Schema obligations	127
	7.4	Database implementation mechanisms	128
8	**Impact on SSADM techniques**		**133**
	8.1	Introduction	133
	8.2	Impact on SSADM steps	133
A	**The System Development Template and the 3-schema Specification Architecture**		**135**
	A.1	Introduction	135
	A.2	The System Development Template	135
	A.3	The 3-schema Specification Architecture	136
	A.4	Main area of impact of OO approach in this volume	137
B	**Specification for Contractor case study**		**139**
C	**Revisions to SSADM operation syntax for entity-event modelling**		**151**
D	**SSADM V4 concepts and corresponding OO concepts**		**155**
Bibliography			**157**
Glossary			**159**
Index			**169**

1 Introduction

1.1 Purpose of this volume

The purpose of this volume is to show how to reduce costs on information system (IS) development projects where the Structured Systems Analysis and Design Methodology (SSADM) is to be used, through increased reuse of common SSADM components.

The increased reuse of common SSADM components is achieved through a limited exploitation of object-oriented (OO) concepts within the established SSADM framework. This volume concentrates on achieving reuse in two specific areas:

- reuse in a data model

- reuse in a behavioural model.

Additionally the volume demonstrates how to develop specifications for method-invocation and message-passing.

None of the ideas in this volume rely on the existence of an OO implementation environment. The ideas can all be easily adopted within an SSADM project using conventional database technology. Nevertheless, none of the ideas preclude the use of extended relational or OO database technology during implementation.

Minimal change to SSADM

Those familiar with the full range of OO implementation ideas may be surprised at the relatively modest changes in SSADM. The changes are modest because the approach described in this volume has been shaped by four notions:

- SSADM customers are seeking maximum benefit with minimum change

- the designs produced must be implementable using conventional database technology, as well as OO databases

- the benefits from some OO ideas are restricted to areas such as graphical user interfaces

- SSADM is already amongst the most object-oriented of system development methodologies designed in the 1980s and early 1990s.

Continuity of model

Modelling notations which might form a barrier between systems analysts and systems builders have been avoided where possible and explanation is given of how system modelling notations used during SSADM analysis and design are carried forward to implementation.

1.2 Who should read this volume

The primary intended reader is the experienced SSADM practitioner who has been trained to the level of the SSADM certificate of proficiency and who has considerable practical experience of using SSADM. Readers will also find it helpful if they are familiar with the 3-schema Specification Architecture as described in the Information Systems Engineering (ISE) Library volume *Customising SSADM*.

In an SSADM context, the intended reader is the analyst working at Stage 3 of SSADM and specifying the Logical Data Model and Effect Correspondence Diagrams (ECDs). In contexts other than SSADM, the intended reader is the information system analyst who has to specify a data model for database builders and database update processes for programmers.

A secondary audience is the designer or program specifier who needs object-oriented alternatives to procedural program specification techniques (such as those in Stage 5 of SSADM).

The volume is not aimed at the analyst specifying a graphical user interface or the designer using a User Interface Management Systems (UIMS). The use of OO ideas in these areas is well established (see for example ISE Library volume *SSADM and GUI Design: A Project Manager's Guide*). The guidance in this volume is perfectly compatible with the approaches used in these areas.

1.3 Status of this volume

CCTA is pleased to present this guidance as part of the ISE Library. However, anyone intending to put the guidance into practice should note that guidance, in the areas covered by this volume, is still evolving and maturing.

1.4 Associated publications

This publication is one of a number of related volumes in the ISE Library:

- *Customising SSADM* – describes the 3-schema Specification Architecture and the basic principles which must be observed in any customisation of SSADM

- *Accelerated SSADM* – describes ways, other than those explained in this volume, of reducing IS development time and cost

- *Application Partitioning and Integration with SSADM* – describes how large systems may be partitioned and subsequently integrated, as may be required for either serial or parallel application development

- *SSADM and GUI Design: A Project Manager's Guide* – written for managers of SSADM projects which are to employ Graphical User Interface (GUI) technology.

1.5 Structure of this volume

Chapter 2 summarises the key OO concepts and their benefits. It is primarily intended for those unfamiliar with OO approaches and terminology. However, even those who are comfortable with OO concepts are recommended to read this chapter as they may find some definitions have been accorded unfamiliar nuances.

Chapter 3 describes the role of OO concepts within SSADM and introduces some of the main concepts described in further detail in Chapters 4 and 5.

Chapter 4 describes extensions to data modelling in SSADM that facilitate reuse.

Chapter 5 describes extensions to process modelling in SSADM that facilitate reuse.

Chapter 6 shows how SSADM techniques may be modified to develop object-oriented specifications for method invocation and message-passing. These specification techniques may be used where a conventional relational database and programming language are to be used during implementation.

Alternatively, they could be required where an extended relational database or even object-oriented database is to be used during implementation.

Chapter 7 briefly considers issues connected with External and Internal Design.

Chapter 8 outlines the impact on SSADM techniques of the extensions described in this volume.

Annex A provides an overview of the System Development Template and the 3-schema Specification Architecture.

Annex B provides specification material for the Contractors case study not included in the main body of the volume.

Annex C describes the modifications to SSADM operation syntax.

Annex D summarises the correspondences between object-oriented and SSADM concepts.

2 OO concepts

2.1 Introduction

This chapter summarises the key OO concepts, their benefits and application. It is primarily intended for those unfamiliar with OO approaches and terminology. However, even those who are comfortable with OO concepts are recommended to read this chapter as they may find some definitions have been accorded unfamiliar nuances.

The chapter is sub-divided into:

- classes and objects (Section 2.2)
- encapsulation (Section 2.3)
- class hierarchies and inheritance (Section 2.4)
- message-passing (Section 2.5)
- methods and polymorphism (Section 2.6)
- benefits of OO concepts (Section 2.7)
- applicability of OO concepts to the development of information-based systems (Section 2.8).

2.2 Classes and objects

A class is a generalised description of a set of object instances which share the same properties. There are two basic kinds of property which the class must describe, 'attributes' and 'methods'. To completely define a class, we must define its data view (all its attributes) and its process view (all the methods which operate upon those attributes). For example, consider the class of objects called 'Task'. Figure 2.1 shows the methods and attributes for 'Task'.

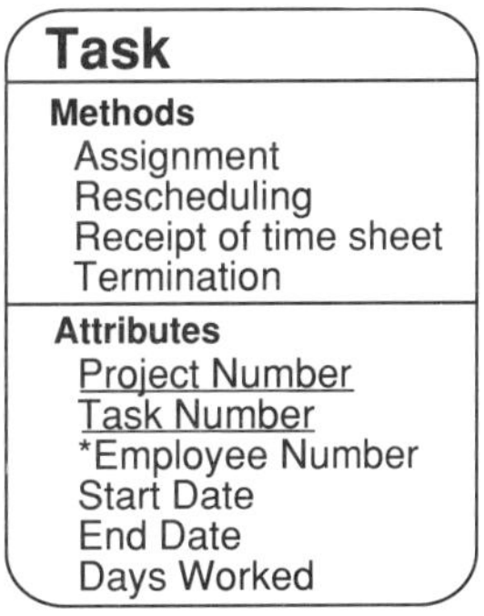

Task
Methods
Assignment
Rescheduling
Receipt of time sheet
Termination
Attributes
Project Number
Task Number
*Employee Number
Start Date
End Date
Days Worked

Figure 2.1: Attributes and methods for class 'Task'

Note that the attributes in Figure 2.1 include a compound key (Project Number and Task Number) denoted by underlining which will act as a unique identifier for each Task instance and a foreign key (Employee Number) denoted by an asterisk.

Methods

A method is a processing routine which belongs to an object; it acts on an input message, processes the attributes and produces a response. Specifying an object's methods is relatively difficult, as is defining methods at the 'right' level of reusability. This volume provides an explanation of how both of these things can be done.

2.3 Encapsulation

Probably the most fundamental object-oriented concept is that the data and processing perspectives of an object are inseparable. This is called 'encapsulation'. Encapsulation means that the processing routines which access the data attributes of an object are somehow enclosed with that data.

Encapsulation is related to the more general idea of 'information hiding'. There is effectively a boundary around an object which separates the processing of the object from its interface to other objects. The only way to get at the data of an object is through its processing routines.

Suppose a particular class of objects has ten properties, four attributes and six methods. Figure 2.2 shows that for each object in the class only the class methods are able to access the attributes of that object and that the methods themselves are only accessible via the object interface.

2.3.1 Encapsulation in system specification

Encapsulation can be represented during system specification by showing all the processes which enquire upon or change data along with that data. We discuss this further in Section 3.6. Encapsulation in system specification is fundamental to the process modelling techniques described in Chapter 5.

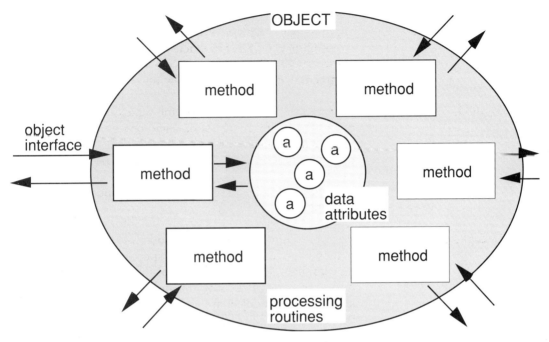

Figure 2.2: Object methods and attributes are encapsulated

One of the main ideas behind object-oriented analysis and design methods is that systems should be constructed by specifying both the data-oriented and process-oriented views of the objects in the system in co-ordination.

2.3.2 Encapsulation in system implementation

Encapsulation can be implemented in objects in a system to control the way they behave at run-time. Encapsulation has the effect that the data items in the data area of an object can only be created or changed during calls to the object. These data items cannot be created or changed between calls to the object, by any other object. An object cannot look directly at the data area of another; but it can indirectly, by sending a message to another asking it for information.

Encapsulation in system implementation usually requires an OO environment, but it is possible to simulate encapsulation in a conventional environment. Implementation issues are briefly discussed in Chapter 7, however, this volume is mainly concerned with improving techniques for system specification.

2.4 Class hierarchies and inheritance

A class hierarchy is a hierarchy of classes in which super-type classes are related to one or more sub-type classes and sub-type classes are related to one super-type class. See Figure 2.3 for a simplified version of a class hierarchy used in Chapter 4. In Figure 2.3 super-type to sub-type relationships are represented by exclusion arcs across one-to-one relationships.

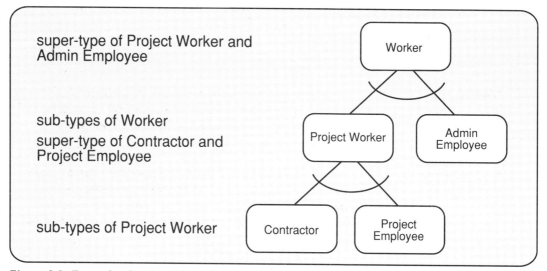

Figure 2.3: Example of a class hierarchy

Inheritance	Within a class hierarchy, a sub-type class is assumed to have all the properties of its super-types (in addition to its own). Super-types record properties common to their sub-types. Sub-types are said to inherit from their super-types.
Multiple inheritance	Sometimes it is permitted for sub-type classes in a class hierarchy to be directly related to more than one super-type class and to inherit the properties of all the super-type classes to which they are related. This type of inheritance is called multiple inheritance.
Inheritance as an implementation mechanism	Inheritance is a specific implementation mechanism by which an OO environment can avoid storing instances of super-type classes. The OO environment implements an object instance as the lowest sub-type it can be matched to, extended with properties inherited from its super-types.

The idea of the inheritance mechanism is less important in systems specification than the basic idea of a class hierarchy.

2.5 Message-passing

When an event or enquiry causes information in the system to be read or updated, any objects involved in the read or update communicate by the exchange of messages.

2.6 Methods and polymorphism

We now introduce three methods-related concepts:

- method bodies
- preconditions
- polymorphism.

Method body

The body of a method is a procedure composed of method body steps, or distinct executable operations and conditions. The specification of method bodies is considered in Chapter 5.

Preconditions

The preconditions of a method define the state an object must be in for that method to work correctly. Usually they can be specified in terms of valid prior states and are coded at the start of a method body as a kind of fail test or quit condition.

Polymorphism

Polymorphism occurs where two or more classes of object respond to the same message by executing different methods. The resulting exhibited behaviour may be different (for example, as a result of a message containing an arithmetic operator being sent to two different classes) or similar (for example, as a result of the message 'rotate' being sent to 2-dimensional and 3-dimensional objects). Polymorphic methods are relevant at the programming level, for example to the design of production software, but are rarely relevant in information system design.

2.7 Benefits and application of OO concepts

Many benefits are claimed for OO including:

- easier maintenance of systems developed using OO
    - this can be partly attributed to encapsulation organising related concepts into groups and thereby making them easier to find during maintenance. It also helps ensure that changes to

the internals of a particular class of objects are isolated and do not affect other classes

– it can also be partly attributed to the practice of naming classes, objects, methods and attributes from analysis through to implementation after things in the system being modelled

- the computer system more directly modelling the system under investigation

– this is especially true for systems which mainly consist of components interacting by the exchange of messages and where most data is short-lived. For example simulation systems, control systems, computer aided manufacturing and real-time systems

– again this benefit can be partly attributed to the practice of naming things in the computer system after things in the system being modelled

- more rapid system development

– this is especially the case for systems which have deep class hierarchies. For reasons discussed in Chapter 4, most information-based systems do not have deep class hierarchies. Examples of types of system which do include those with a strong graphical element such as windows-based operating systems or computer aided design

– large systems, especially, can be partitioned for incremental delivery and then at a later stage the larger system can be assembled from the smaller ones.

The problem then is how to integrate these smaller applications without a large amount of redevelopment effort. ISE Library volume *Application Partitioning and Integration with SSADM* provides practical guidance on this application partitioning and subsequent integration.

However, leaving aside the potential benefits due to eased system maintenance, where there is still little practical experience, the principal benefit of OO concepts which can be realised now is reuse.

Reuse

Businesses can reduce costs by increasing reuse. Reuse of components should save time, both during development and testing. However, additional time will be needed up front both in planning a 'programme' of projects to capitalise on the reuse potential, and in ensuring during systems analysis that reusable components are indeed reused.

There are of course other ways of achieving reuse that are not object-oriented. These tend to involve standardisation and bottom-up design. For further information refer to ISE Library volumes *An Introduction to Reuse* and *Managing Reuse*.

2.8 Applicability of OO concepts to the development of information-based systems

OO programming languages are popular for writing compilers, CASE tools and GUI software, but are not generally suitable for the kind of system for which SSADM is primarily designed, which maintains large volumes of persistent data. The need for storage and maintenance of a permanent data structure means that OO programming ideas must be tailored for successful use in SSADM.

The volume draws on evidence and experience from developing information systems rather than production software. This volume warns readers where there are practical difficulties and limitations. The reader should not expect massive savings from the use of OO concepts. Above all OO concepts should not be seen as a panacea.

The techniques in this volume integrate the ideas of data modelling, entity process modelling and object-oriented programming. They can be applied with current technology and supported by current CASE tools. They do not require the use of an object-oriented database management system.

3 OO concepts in SSADM

3.1 Introduction

This chapter describes the role of OO concepts in SSADM V4. The chapter is divided into the following sections:

- OO paradigms (Section 3.2)

- role of the OO paradigms in SSADM V4 (Section 3.3)

- implementation independence of Conceptual Model (Section 3.4)

- objects as 'long-running' processes (Section 3.5)

- encapsulation in SSADM (Section 3.6)

- composite objects, composite methods and aggregate objects (Section 3.7)

- class hierarchies in SSADM (Section 3.8).

3.2 OO paradigms

OO is based on the assumption that there are objects common to different programs, or even systems, which can be defined independently of those programs or systems, and reused over and over again. Reuse is the primary benefit which those in the information systems community are looking for.

Object orientation is a proven technology at the programming level which brings clear benefits in terms of reuse. However, there are significant differences in the way in which OO has to operate at the information systems level. These differences are represented in the two separate paradigms of OO design – method-structuring and control-structuring.

Method-structuring

Method-structuring evolved in the development of Smalltalk, and is now embodied in other languages such as C++.

The code is separated into a number of loosely coupled 'methods' which are invoked by message-passing.

Objects are classified and organised into a hierarchy with methods definable at any level in the hierarchy.

OO programming based on method-structuring has been demonstrably successful in the development of GUIs and system software. This success relies on:

- relatively small numbers of objects

- simple mainly hierarchical object models; little co-ordination needed between objects in different type hierarchies

- most objects having low or no requirements for persistence

- relatively few object types visible to users.

Control-structuring
Control-structuring evolved via Jackson structured programming and is now embodied in entity-event modelling in SSADM. The control-structured paradigm deals with:

- indefinitely large numbers of object instances, whose integrity needs to be maintained

- complex non-hierarchical object models, requiring co-ordination across object types

- data objects persisting over a long period

- many object types visible to users, leading to a wide range of input types.

The control-structured approach is well suited to specification of systems that have to manage databases, although additional techniques are needed which have not featured strongly in method-centred OO experience.

3.3 Role of the OO paradigms in SSADM

In order to properly understand this section it is necessary to first understand the concepts of the 3-schema Specification Architecture and the System Development Template. Any reader unfamiliar with these concepts or wishing to refresh their memory is recommended to refer to Annex A.

3.3.1 Role of the two paradigms in SSADM

Both the OO approaches described in Section 3.2 fit into SSADM. Figure 3.1 shows three categories (resulting from the 3-schema Specification Architecture) of processing to be implemented:

- user interface processes (from the External Design) – candidates for the method-structured treatment

- database update and enquiry processes (from the Conceptual Model) – candidates for the control-structured treatment

- program-data interface processes (from the Internal Design) – invoked by methods in the database update process to obtain entity data from the database, and return it there after it has been changed.

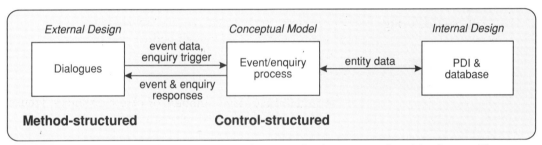

Figure 3.1: Relationship of the two OO paradigms to the three categories of implementable processes

3.3.2 Impact of changes to SSADM in context of SDT

SSADM concentrates on the specification of the Conceptual Model. Therefore this volume concentrates on extensions to conceptual modelling techniques which help with reusability in both data and processing. Figure 3.2 shows these changes in the context of the System Development Template.

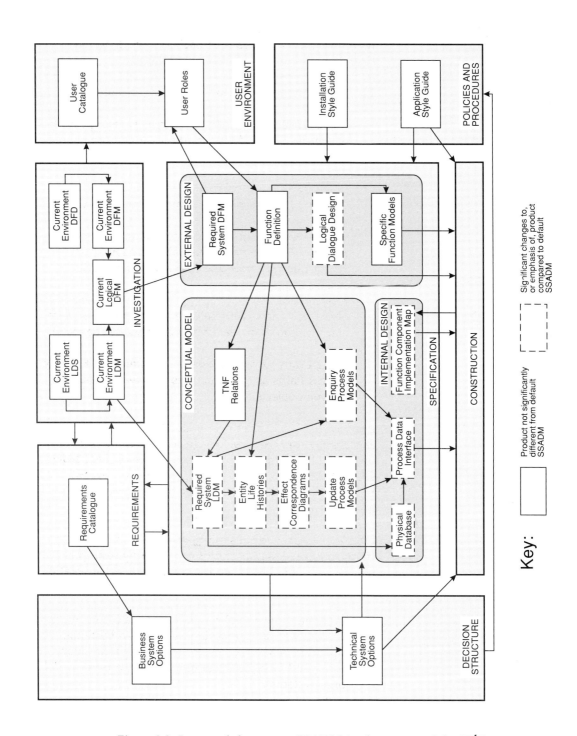

Figure 3.2: Impact of changes on SSADM in the context of the SDT

Chapter 3
OO concepts in SSADM

Conceptual Model
: The extensions to conceptual modelling are compatible with method-structured approaches in the External Design.

  The extensions to conceptual modelling include:

  - Logical Data Model: type hierarchies, and splitting of entities into 'aspects', to discover common behaviour

  - in Entity Life Histories, identifying 'super-events' – common processes used by more than one event

  - as an alternative to Update and Enquiry Process Models, developing method specifications from Entity Life Histories and Enquiry Access Paths

  - specification of an 'event manager' (a transient object) for each event, to co-ordinate the invocation and outputs of all the objects affected by the event

  - using Effect Correspondence Diagrams and Enquiry Access Paths as the structures for message-passing between objects.

External Design
: External Design is briefly mentioned in Chapter 7. We assume that the External Design will be developed with one of the method-structured approaches derived from OO programming experience, as encouraged by ISE Library volume *SSADM and GUI Design: A Project Manager's Guide.*

Internal Design
: Internal Design is dealt with briefly in Chapter 7. The (fairly informal) syntax used for processing examples would map very easily on to a relational implementation with an SQL-based interface.

  However, the Conceptual Model processes would work equally well with an object database system (although the implementor would usually have to do more work on locking and two-phase commit than with most relational systems).

3.3.3 Technical architecture

Perhaps the best compromise, between the ideal technical environment and what is currently well-supported in technology, is a distributed client/server architecture. A first-cut design could be as shown in Figure 3.3.

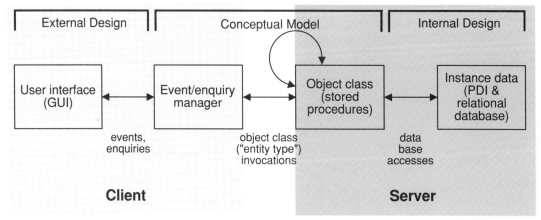

Figure 3.3: Possible client/server architecture for implementation

3.4 Implementation independence of Conceptual Model

Methodologies like SSADM are largely about building a machine-independent systems specification, or 'model' of the system requirements. The Conceptual Model developed at Stage 3 of SSADM must be a generalisation from successful implementation mechanisms (see Figure 3.4); it must not dictate the choice of implementation mechanism.

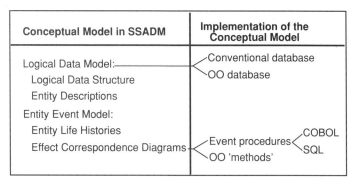

Figure 3.4: The Conceptual Model must be a generalisation of implementation mechanisms

In other words, the LDM should be unaffected by whether the database management system is 'relational' or 'object-oriented' and the ECDs should be unaffected by whether the programs are written in COBOL, C++ or SQL.

What the SSADM practitioner needs help in is the discovery, design and representation of:

- encapsulated entity processing routines (or 'methods')

- common components (or super-types in class hierarchies).

3.5 Objects as 'long-running' processes

Meyer (1988) says, 'by introducing a state and operations on this state, we make the [object] specification richer' and 'the view of objects as state machines reflects [object] types which are more operational, but this in no way makes them any less abstract'.

We can view an object as a 'long-running' process and attributes of an object as the working-storage of this long-running process.

Taking the long-running process view further, we can define an object as a finite state machine. This means that given the current state of an object and the input message it is to process, it is possible to specify or predict the new state of the object and any output message it will produce. Each object must be in one of a finite number of states. All the possible states of an object, and the transitions between them, can be specified.

The state-transition view of objects is an important OO idea that is firmly embedded in SSADM through the use of Entity Life Histories.

3.6 Encapsulation in SSADM

Encapsulation features strongly in SSADM through ELHs which are used to specify all the update processes of the entities in the LDM. But where are the 'methods' in SSADM? Does a method correspond to an operation on an ELH? the effect of an event? or something else?

Initially, we assume that as far as one entity is concerned, each event invokes one method and each method is invoked by one event. We name methods after the events which trigger them. We will refine this simple view in Chapter 5.

3.6.1 Showing object properties on the LDM

It is sometimes helpful when working on the detail of a small part of an LDM, especially where the LDM is being developed using a CASE tool, to expand an entity box, as in Figure 3.5, to display the names of the entity's attributes, or the events or methods which affect that entity.

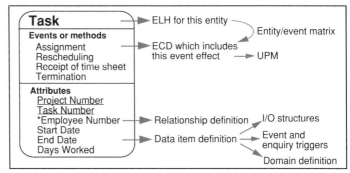

Figure 3.5: SSADM backing documentation for an entity

Since the 1970s, attempts have been made to extend data modelling notations like that used in the LDM to specify a system's business rules more completely. There is a limit to what can be achieved, without making the LDM too large and confusing to be useful. So except for the smallest LDMs, each entity will require backing documentation, preferably maintained along with the LDM by a CASE tool.

3.6.2 Developing the behavioural model

Listing the names of events or methods affecting an entity along with that entity is only a small step towards object-orientation. More important than whether the methods are shown on the LDM is how we:

- discover the 'right' set of events or methods

- name them

- specify their workings

- specify the message-passing between objects
- specify and maximise reuse of methods.

In other words, we need a systematic way of developing the behavioural model. The relevant techniques are outlined in Chapter 5.

3.6.3 Reusable methods

One event can be input in many different contexts. So the process for an event in the Conceptual Model can be reused in different parts of the External Design. The *SSADM V4 Reference Manuals* lack a way of identifying further opportunities for reuse within an event. This is rectified in Chapter 5.

3.6.4 Composite methods

For each event in the ELHs, we can produce an ECD. This specifies the co-ordination of processing routines, in different objects, which are triggered by one event. Given an LDM and a set of ELHs, many of the ECDs can be generated automatically. An ECD can be thought of as a composite of processing routines triggered by one event or one super-event, and may be regarded as a composite method. This is described in more detail in Section 3.7.2.

3.7 Composite objects and methods, and aggregate objects

Most OO authorities draw a distinction between composition and aggregation. However, for our purposes it is convenient to define three different forms of composition and aggregation:

- composite object
- composite method
- aggregate object.

3.7.1 Composite objects in data modelling

We define a **composite object** as a set of parallel or nested objects sharing or inheriting the same identity. A composite object need not be in second or third normal form and may include lists or repeating data.

In External Design there are composite objects such as windows and reference lists and GUI software design needs to be able to handle composite objects. But where are composite objects useful in the Logical Data Model?

The prime possibility is that a composite object should be remodelled as a master entity with a set of one or more detail entities, where the whole shares the same identity.

For example, we could think of an Order as a composite of Order Items. This is not particularly useful because Order Items usually need to be identified individually during processing so that they can be added, altered or cancelled one-by-one. It is true that the delete operation can be propagated from the composite object to its components, but this operation is to do with database management rather than the real-world model. If events in the Conceptual Model such as 'Delivery' and 'Invoice' trigger the propagation of methods from the Order to its Order Items, this is better modelled in the Effect Correspondence Diagrams (ECDs) than in the Logical Data Model.

3.7.2 Composite methods in process modelling

Remember the purpose of defining a composite object is to model a situation where a method on a composite object is propagated to its component objects. For example, if a Polygon is a composite object made up of several Point objects, then the operation 'Move Polygon' must trigger several 'Move Point' operations.

Does propagation occur in information systems?

Yes, but for most events in an information system, the method in the composite object would be different from the methods in the component objects. The method to cancel an Order must trigger a method in each Order Item to cancel it, but it will for example set the state variable to a different value, so simple propagation is not enough.

What is the equivalent idea in information systems?

Composite objects do appear in SSADM's conceptual modelling techniques, but in the ECDs rather than the LDM. Every event is a transient composite object and fires a composite method. We define a **composite method** as a set of methods operating on one or more object types, all triggered by the same event or super-event. We record a composite method in the domain of process modelling (in an ECD) rather than data modelling.

Why not show composite methods in the LDM?

There is a mismatch of composite method and object lifetimes. Composition features strongly in OO environments which have been developed in the context of graphical user interfaces, where all the objects on the screen live about the same length of time. It is not hard to imagine both the objects and their composite methods being specified in one LDM. Unfortunately, in information systems, composition is usually very short-lived compared with the long life of the stored data. To show all the necessary short-lived compositions in the LDM, it would be necessary to include each event in the LDM, perhaps as a box surrounding the objects which it effects. A composite created to enable propagation of methods needs to last only for the duration of the event which triggers the methods. For information systems, it is better to model such composition in the specification of events.

3.7.3 Aggregate objects in data modelling

We define an **aggregate object** as a restricted type of composite object owning a set of parallel objects all dependent on the same primary key. So an aggregation is fully normalised and cannot include lists or repeating data. There are three ways such an aggregation might be implemented:

- first, we could turn the attributes into key only entities, or operational masters as described in SSADM Version 3

- second, we could implement the aggregate object as one relational database record. In this case, physical database management operations like read, write and delete record will be propagated from the aggregate to all its components; but note that these database management operations are not part of the real world model

- third and most interestingly, we could pair off each attribute with the key of the main object, so each attribute becomes a **parallel aspect**. The ISE Library volumes *Application Partitioning and Integration with SSADM* and *Distributed Systems: Application Development* describe the practical uses of this approach.

3.8 Class hierarchies in SSADM

True reuse of common components implies a class hierarchy. Class hierarchies are not strongly featured in the SSADM manuals, and there is room to strengthen this aspect of SSADM.

Ideally, we need a way of defining a class hierarchy which does not limit it to data modelling. A class hierarchy can be defined as a structure in which super-type nodes are divided into sub-type nodes. The super-type node provides a place to record properties common to all its sub-types.

The main benefit that analysts should gain from discovering or specifying class hierarchies is less redundancy in the system specification by greater reuse of common components. Several types of class hierarchy can be found in SSADM:

- common domains (Section 3.8.1)
- common data groups (Section 3.8.2)
- super-type entities (Section 3.8.3)
- super-events (Section 3.8.4).

3.8.1 Common domains in data item definition

Where two or more data items (whether attributes of entities or items in an I/O structure) share the same definition, then it is sensible to create a common domain or data type to record this.

Where insufficient data types are provided by the software implementation language, designers can choose to define their own data types, specific to the system being designed.

In practice, most data items require only one level of domain definition between data type and data item. But some data items can inherit two or more levels of domain definition. Country of Origin is an example:

- Standard data type: Text
- User-defined domain 1:   Country Name
- User-defined domain 2:   European Country
- User-defined data item: Country of Origin

A deep and complex structure of data types or domains could be created, but in practice, multi-level restrictions like this are rare. They are usually defined by combining data dictionary domain definition with a local external domain definition, such as a validation rule connected to a data entry box in an input window.

3.8.2 Common data groups in data flows

Where two or more data flow structures share the same data group, then it is sensible to create a common data group to record this. A deep and complex structure of data groups could be created.

In practice, it is often convenient to define the bottom level, or elementary component, of an I/O structure or data flow in terms of components of the Conceptual Model, so that:

- output structures are defined in terms of entities
- input structures are defined in terms of events and enquiry triggers.

3.8.3 Super-type entities in the LDM

Data types are usually thought of as applying to individual data items, but what about groups of data items, or entity types? Where two or more entity types share data items, then it is possible to create a super-type in the LDM to record these common properties. Whose responsibility is it to define the most generic record type in the example in Figure 3.6?

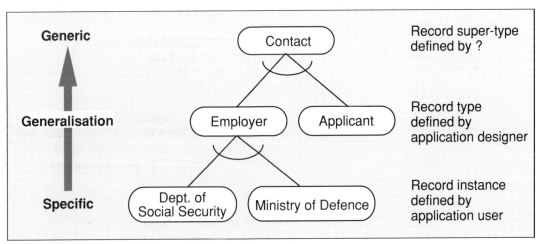

Figure 3.6: Generalisation in an LDM

In Figure 3.6, we define the record types Employer and Applicant. We also define the Contact super-type to accommodate data and processing common to both Employer and Applicant. But if the Contact super-type could be reused in another application, perhaps the job of designing this reusable component would better be done by the database manager.

Practical experience has demonstrated that super-types, such as Person, Organisation and Location, which might be shared between information systems are unlikely to save more than one or two per cent of system development effort. However, there are cases where an application-specific LDM class hierarchy can deliver larger savings.

The identification and use of LDM super-types, to minimise redundant data and processing and maximize reuse, is discussed in detail in Chapter 4.

3.8.4 Super-events in entity-event modelling

Where two or more events share the same effect on an entity, then it is sensible to create a common component (or super-event) to record this in the ELH. See Figure 3.7. One benefit of a super-event occurs where it can be treated as an ordinary event in the behaviour models of related entities, saving redundancy of specification.

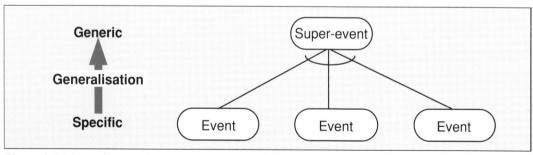

Figure 3.7: Generalisation in entity-event modelling

A second benefit of the super-event concept occurs during process specification. A common processing routine is created for the super-event. Chapter 5 shows that the development of a relatively complex network of common processing routines can be simplified through the use of super-events.

3.8.5 Achieving reuse

In this volume we focus mainly on the discovery, representation and design of entity class hierarchies and event class hierarchies. Both types of hierarchy are aimed at maximising reuse.

As far as achieving reuse is concerned entity super-types form only a small percentage of a typical information system model. Fortunately, super-events are more common.

4 Data modelling extensions

4.1 Introduction

This chapter describes how to include reusable components in Logical Data Models. It covers the use of mutually exclusive relationships, entity class hierarchies, and generalised reusable entities.

4.1.1 Event class hierarchies yield more reuse

The main OO concept used in this chapter is the notion of a class hierarchy. However, we should not expect too much in the way of reusable components from class hierarchies of entities. In general there is more reuse to be achieved by developing *event* class hierarchies (described in Chapter 5) than by developing *entity* class hierarchies. Nevertheless the benefits of entity class hierarchies are tangible and worthwhile, and can be achieved with no significant increase in the data modelling effort that might be expected if the potential for reuse was ignored.

4.1.2 Structure of this chapter

This chapter is sub-divided into five main categories:

- concepts and notation
    - class hierarchies (Section 4.2)
    - aspects (Section 4.3)
- achieving reuse in data modelling (Sections 4.4–4.7)
    - discovering class hierarchies (Section 4.4)
    - development of the Logical Data Model (Section 4.5)
    - using aspects to prepare the LDM for detailed development (Section 4.6)
    - example of achieving reuse (Section 4.7)
- effect of object behaviour on reuse (Section 4.8)
- limits to reuse (Section 4.9)
- validating the LDM by entity-event modelling (Section 4.10).

Section 4.2 refreshes and refines some facets of SSADM data modelling practice that are essential to an understanding of reuse opportunities. It covers the basic ideas of class hierarchies, sub- and super-types, and mutual exclusion.

Section 4.3 introduces entity aspects for modelling distinct behaviours in a single entity type.

Sections 4.4 to 4.7 describe, using a case study, how class hierarchies should be developed to achieve reuse. This guidance is concerned with reuse *within* a single system.

Section 4.8 explains the role of object behaviour in determining whether there is real reuse benefit to be had from entity class hierarchies.

Section 4.9 describes why reuse of super-type entities *between* systems is so difficult to achieve in practice.

Section 4.10 explains that entity-event modelling should be used to validate and where necessary to enhance the LDM.

4.1.3   Contractor case study

Many of the examples in this volume are based on versions of the Contractor case study. See Figure 4.1 for the initial version of the LDM.

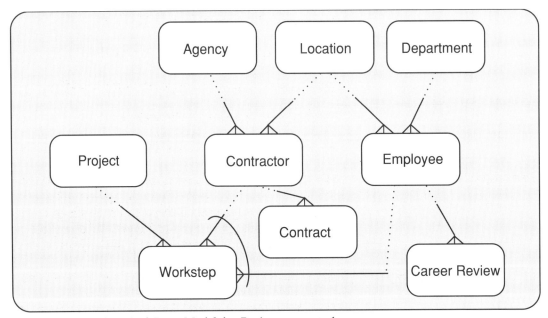

Figure 4.1: Initial Logical Data Model for Projects case study

Chapter 4
Data modelling extensions

4.2 Class hierarchies

The basic notion of a class hierarchy was described in Section 2.4. The ideas of a class hierarchy and inheritance tree are useful in data modelling, but the idea of inheritance as an implementation mechanism is not. The analyst building an LDM need not, and should not, be concerned with whether the database designer will employ inheritance, or one of the alternative implementation mechanisms.

A more important issue is why *during analysis* we may find it advisable to alter what at first seems to be a class hierarchy by either rolling the super-type down into distinct sub-types, or rolling up the sub-types into their super-type, or converting a class hierarchy into an aggregation. This is quite separate from what the database designer may choose to do during implementation.

In this section we will consider:

- sub-types and super-types (Section 4.2.1)

- mandatory sub-types (Section 4.2.2)

- optional sub-types (Section 4.2.3)

- hierarchies of sub-types (Section 4.2.4)

- orthogonal (parallel sub-types) (Section 4.2.5)

- multiple inheritance (Section 4.2.6)

- combining class hierarchies (Section 4.2.7).

4.2.1 Sub-types and super-types

A class hierarchy models two or more different kinds of entity in two views – a super-type and a set of sub-types. The super-type models the common behaviour (data and processing) of the entity types; the sub-types model the differing behaviours of the entity types.

4.2.2 Mandatory sub-types

To show such an entity class hierarchy on a data model, we can draw an exclusion arc across one-to-one relationships. The arc makes the relationships optional at the super-type end. In Figure 4.2, a Location must be a

37

company Office or an employee's Home (and nothing else); a Worker must be an Employer or a Contractor.

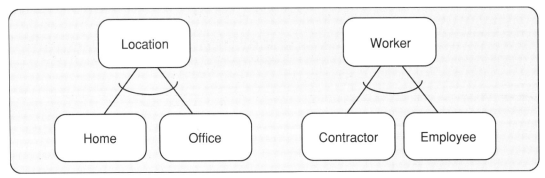

Figure 4.2: Mandatory sub-types

4.2.3 Optional sub-types

To show the super-type does not have to be one of the sub-types, we can draw the relationships as optional at the super-type end as in Figure 4.3.

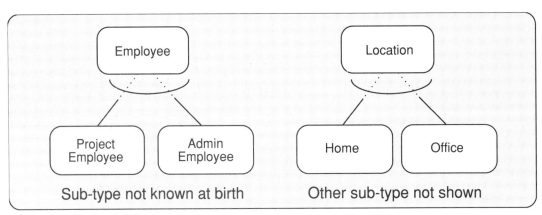

Figure 4.3: Optional sub-types

This can have two meanings. First, it can mean the sub-type is not yet known; for example an Employee will eventually be designated either a Project Employee (an employee who is qualified to work on projects) or an Admin Employee, but can be recorded in the system before it is known which kind of employee he is.

Second, it can mean there is another sub-type not shown, perhaps because it has no additional properties. In Figure 4.3, Locations other than company Offices and employees' Homes could be recorded in the system.

4.2.4 Hierarchies of sub-types

Hierarchies of sub-types may be defined indefinitely deep, (although in practice hierarchies of more than two levels seem to be fairly rare).

In Figure 4.4, every Worker is either a Project Worker or an Admin Employee; every Project Worker is either a Contractor or a Project Employee.

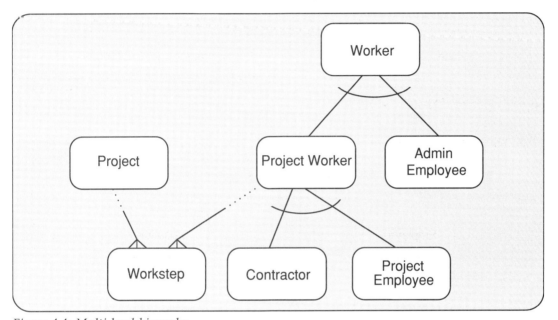

Figure 4.4: Multi-level hierarchy

Note that if we had looked only at workers, the hierarchy could have been drawn differently:

- as three sub-types of Worker

- Worker sub-typed as Contractor and Employee; Employee sub-typed as Project Employee and Admin Employee.

The hierarchy has been drawn as in Figure 4.4 to model the business rules:

- the relationship of a Workstep with a Worker assigned to carry it out is the same, whether the Worker is a Contractor or an Employee

- Admin Employees are not assigned to Project Worksteps.

4.2.5 Orthogonal (parallel) sub-types

The same entity type may be sub-typed in different, independent ways. For example, Figure 4.5 shows that Workers may be Tele Workers (working from home) or Office-based Workers, regardless of whether they are Project Workers or Admin Employees.

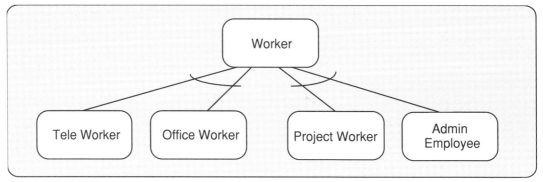

Figure 4.5: Orthogonal sub-types

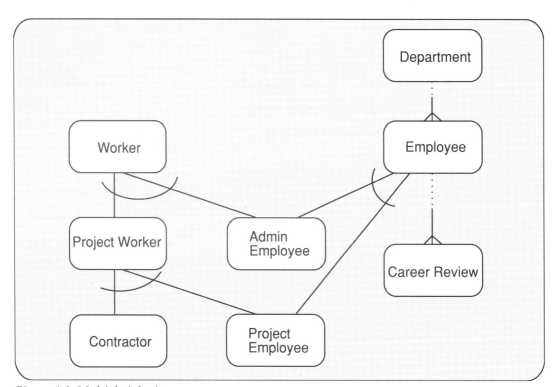

Figure 4.6: Multiple inheritance

4.2.6 Multiple inheritance   Some sub-types may belong to (and inherit properties of) more than one super-type. For example, in Figure 4.6, Admin Employee and Project Employee, as well as being sub-types of Worker, also have common behaviour as a result of belonging to Departments and having Career Reviews; this is modelled by the Employee super-type.

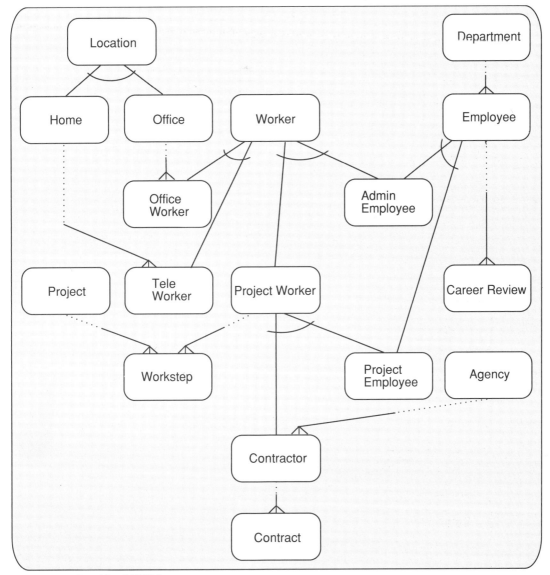

Figure 4.7: Combined LDM

4.2.7 Combining class hierarchies

We can combine the models we have looked at so far (including the initial LDM of Figure 4.1) to give a more precise picture of the case study as in Figure 4.7.

Same business rules – different LDMs

Note that the same real world relationships can be represented by LDMs which look different. We need to learn to recognise when different diagrams are equivalent, that is express exactly the same set of business rules.

For example, in Figure 4.8 an entity class hierarchy is shown by an exclusion arc drawn across one-to-one relationships.

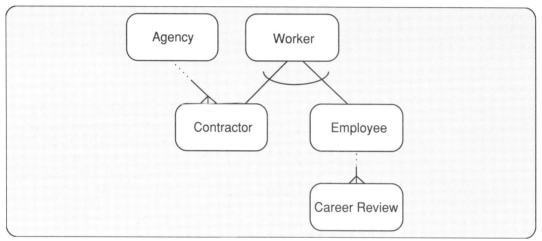

Figure 4.8: Explicit class hierarchy

We can show the same thing by drawing exclusion arcs across one-to-many relationships, as in Figure 4.9.

In Figure 4.9, one might say the sub-types have been 'rolled up' into their super-type. But although the sub-types are not shown explicitly, the exclusion arc shows they are still present. They will still appear elsewhere in the system specification, in the form of options in processing routines.

We need to recognise these two kinds of model can be exactly equivalent. Note that the entity-event model is not affected. The same entity-event model underlies any presentation of an LDM. Behind either of the LDMs in Figures 4.8 and 4.9, there will be an ELH for Worker with a high-level selection where the two options are Contractor and Employee.

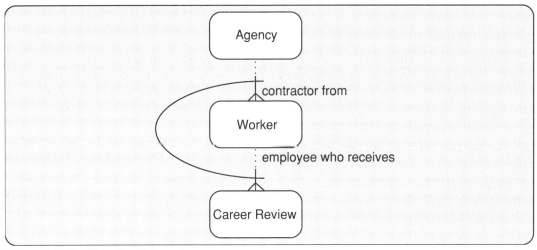

Figure 4.9: Implicit class hierarchy

Same LDM – different business rules

The same LDM can represent different things, that is, be a front for different entity-event models. We have to live with some ambiguity because it is not possible to express all possible semantics in an LDM. In general, an LDM does not tell the whole story; for this we have to look at the data and processing detail suppressed behind the LDM.

In order to avoid any possible confusion over the notation for an entity class hierarchy we can further distinguish entity class hierarchies by the convention of labelling all lines between boxes with a relationship role name *except* within a class hierarchy, where the line must always be an 'is a' relationship.

4.3 Aspects

A class hierarchy provides a way of modelling the common behaviour of different entity types. Another useful technique is that of modelling multiple, parallel behaviours of a single entity type. We can do this by defining **aspects**.

An aspect is an application's view of a real-world entity type, modelling part of its behaviour. What we usually call 'an entity' on an LDM is really an aspect, since it represents only a partial view of the real-world entity type. In most cases a real-world entity type has only one LDM aspect – there is no confusion in referring to 'LDM entities'. We need to distinguish aspects only when there are multiple views of the same entity that have to be co-ordinated in some way.

An entity type has a 'basic existence' aspect, represented by its primary key (and, if its key is hierarchical, the corresponding master relationship). Other aspects are constrained by the birth and death of the basic aspect.

In theory, every other attribute and relationship could be modelled as a separate aspect. In practice, this is rarely useful. Our approach is to keep aspects as large as possible, and split them only when the real-world entity has distinct behaviours that need to be modelled separately – for example, when it has to be represented in different applications or has parallel lives within an application. Shared attributes can reside in the basic aspect.

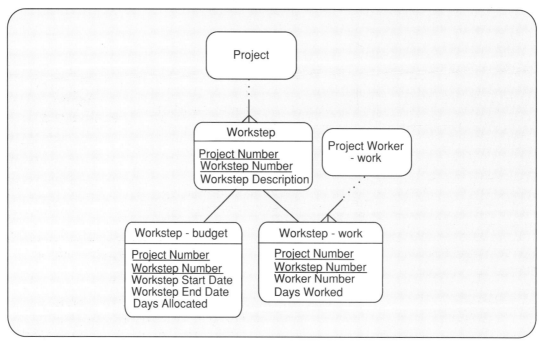

Figure 4.10: Workstep - budget and Workstep - work aspects tied to basic aspect for Workstep

In theory, an entity should not have two aspects; it should have one, or three or more. If it has one behaviour that needs to be separated from its basic existence, it should have another behaviour that is distinct from its basic aspect. For example, Workstep could have parallel lives for budgeting and work. Both of these should be tied to Workstep's basic aspect, which models a workstep's existence as part of a project. See Figure 4.10.

In many cases in practice, basic existence can be merged with another aspect without causing any problems. For example, Workstep and Workstep-budget can be merged.

4.3.1 Aspects and inheritance

Using aspects gives a consistent way of modelling entity behaviour from corporate model to entity within subsystem. For example, Figure 4.11 illustrates a possible hierarchy of Client aspects for a services company.

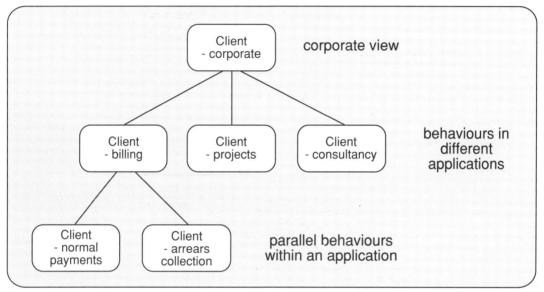

Figure 4.11: Aspects provide a consistent way of modelling behaviour from corporate model to entity

4.3.2 Aspects and aggregation

It is easy to visualise how an aspect within a project could inherit the data and processes of its corporate entity, or how, say, the normal payments aspect of Client could inherit the data and processes of the billing aspect. However, pictures like the one in Figure 4.11 must not be confused with sub-type hierarchies (class hierarchies in OO). One-to-one relationships 'coming down' a sub-type hierarchy are mutually exclusive.

All aspects of a single entity instance can exist concurrently. And (in theory, at least) an entity aspect can invoke any other aspect of the same entity in event or enquiry processing. This is a restricted version of the OO concept of aggregation – the restriction being that all the aspects are views of the same real-world entity. (Aggregation, as it is usually described in an OO context, also accommodates the construction of aggregates of different types of entity.)

4.3.3 Aspects and sub-types

In a sense, defining entity aspects is the opposite of defining entity sub- and super-types, and both are useful in Logical Data Models.

With sub- and super-types we identify *similar behaviour of different real-world entity types*. This behaviour should be modelled within one application. For example, project workers could be employees or contractors. Every instance of project worker would be either an employee or a contractor. We would want to model the 'project worker' behaviour of employees and contractors in the same application as shown in Figure 4.12.

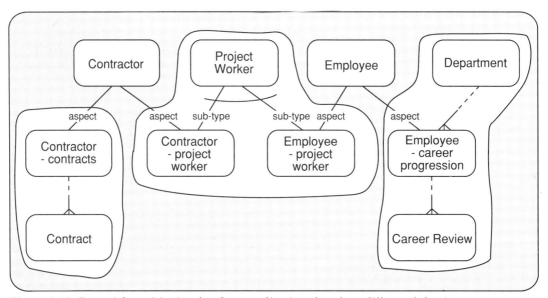

Figure 4.12: Potential partitioning for three applications based on different behaviour

With aspects we identify *different behaviours of the same real-world entity type*. These behaviours can be modelled in different applications. For example, a contractor works on projects and has contracts. There are interactions between the two behaviours that need to be co-ordinated:

- the contractor name and address is needed in both behaviours

- a contractor should not be working on projects unless they have a current contract.

But otherwise the behaviours are largely independent. A contract could span several projects, a new contract could be made part-way through a project, contract terms could be changed at any time. Provided that the interactions are handled, we could model contractor in separate applications for project work and contracts. Similarly, we could model employee in separate applications for project work and career progression. For further exploration of these possibilities see ISE Library volume *Application Partitioning and Integration with SSADM*.

4.4 Discovering class hierarchies

The previous section provided a reminder of some important SSADM data modelling concepts and showed how class hierarchies could be represented. But how are class hierarchies identified?

Class hierarchies are identified by analysing the basic entity types, looking for sub-types and super-types. This analysis can be bottom-up or top-down.

4.4.1 Bottom up generalisation of sub-types into a class hierarchy

Two entity types in the case study benefit from bottom up generalisation. Contractor and Employee have common properties and can be viewed as sub-types of Worker. See Figure 4.13.

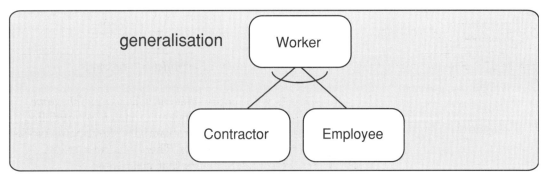

Figure 4.13: Generalisation of Contractor and Employee into Worker

| Keys in generalisation | A generalisation such as Worker is a model optimisation that does not have to be visible to the user. As far as the end-user is concerned the relevant entities are Employee and Contractor. The generalisation of some Employee and Contractor properties into Worker should be hidden inside the system and the notion of a Worker Number (distinct from Employee Number and Contractor Number) is of no relevance to the user. |

Thus, Worker Number becomes the key of 'Worker' and, inside the system, every instance of Employee and Contractor is identified by Worker Number. The user needs to be able to access instances of Employee and Contractor by external, user friendly keys, so Employee Number and Contractor Number are retained as attributes.

4.4.2 Top down specialisation of an entity into sub-types

Employee and Location in Figure 4.1 benefit from top down specialisation. See Figure 4.14. Employee types will be meaningful to the user and should be accessible by, for example, queries on Employee Type.

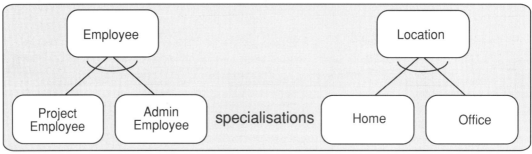

Figure 4.14: Specialisations of Employee and Location into sub-types

| Keys in specialisation | As with generalisation, we could create internal, meaningless identifiers for Employee and Location. However, we also have the option of using the user-friendly keys as the internal keys. |

4.5 Development of the Logical Data Model

As development progresses, we have different concerns about what has to be investigated and modelled in the LDM. There are three major phases:

- understanding and uniquely identifying entities and relationships; we might call this the 'identity' view

- looking for similarities and differences in entities and relationships and creating type hierarchies; we might call this the 'type' view

- creating a detailed, fully-attributed model that will be used as the basis for entity-event modelling (object behaviour modelling); we might call this the 'type' view.

This is not a rigid progression. Experienced modellers tend to transfer seamlessly between these different views of the LDM, documenting class hierarchies and attributes in local areas of the LDM as they discover them. But it is the general sequence in which development occurs.

4.5.1 Identity-based view

The simplest view of an entity type is based on the idea that instances of the entity type are uniquely identifiable by a key. It is reasonable to suggest the analyst would take this simple view at Step 210 of SSADM. For example, Figure 4.15 gives eight types of business related entity. Each would be meaningful to a user for definition of requirements.

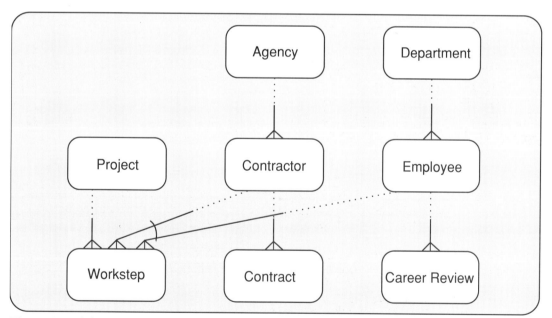

Figure 4.15: 'Identity-oriented' data model

4.5.2 Type-based view

It is reasonable to suggest the analyst should analyse entity class hierarchies, as in Figure 4.16, at Step 320 of SSADM. Although it looks different, this model represents the same problem domain as the data model in Figure 4.15.

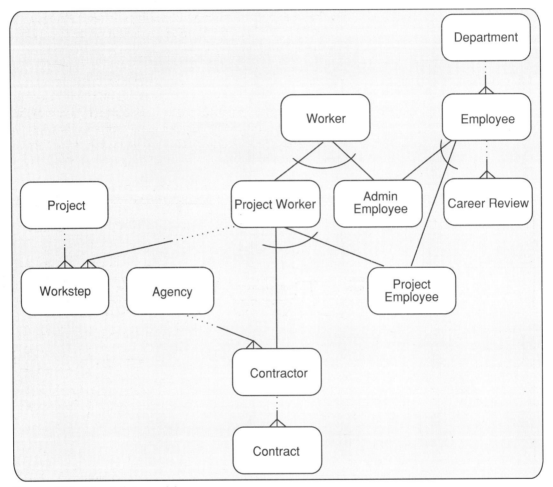

Figure 4.16: 'Type' data model

Four more entity types are shown. Two are sub-types of Employee. They correspond to entities the user would recognise in the business: employees who can be assigned to project worksteps (whose time is specifically charged to projects) and administrative employee (whose time is a general overhead). The user may need access by Employee Type as well as by Employee Number.

Chapter 4
Data modelling extensions

The other two are super-types of Contractor and Project Employee, and Project Worker and Admin Employee. They could be hidden inside the system and the end-user will access Employees and Contractors by employee and contractor identifiers. The end-user should not be presented with new identifiers 'Project Worker Number' and 'Worker Number' in addition to the contractor and employee identifiers.

4.5.3 Object or aspect-based view

Perhaps the least obvious view is that an entity is something which progresses through a defined series of states. However, it is this view of the object data which corresponds best to the object processing specification, so it is reasonable to suggest the analyst should take this view, illustrated in Figure 4.17, at Step 360 of SSADM.

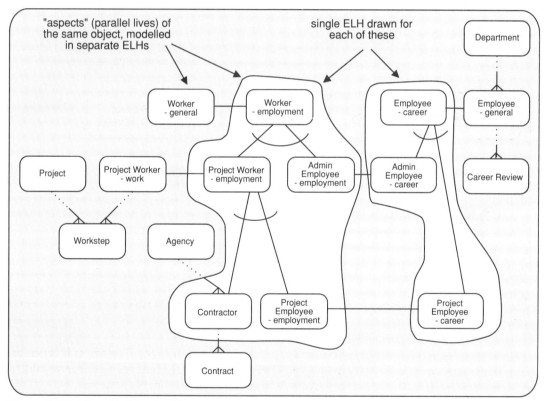

Figure 4.17: 'Aspect' data model

Although it looks different, the model in Figure 4.17 represents exactly the same problem domain as the models in Figures 4.15 and 4.16.

The LDM in Figure 4.17 has been constructed to support easy exploration of object behaviour using entity-event modelling, although it may be simplified in relational data analysis (RDA). It contains three kinds of 'entity':

- entities in the conventional LDM sense, such as Project and Agency

- 'aspects', partial views of entities, such as 'Worker - general' and 'Project Worker - work'

- hierarchies of aspects, Worker - employment and Employee - career.

An ELH is developed for each of them, as described in Chapter 5. They are the things that appear as distinct in their own right in the entity-event model, although the external view to the user will still be 'Contractor' and 'Employee'. The reasons for this approach are discussed in Section 4.6.

4.6 Using aspects to prepare the LDM for detailed development

The introduction of class hierarchies has implications for relational data analysis, entity-event modelling and physical design.

Figure 4.17 showed an object or aspect-based view of the Projects LDM. An aspect such as Worker - general will have attributes common to all types of (real world) worker, for example name and current address, that can be inherited by Contractors, Project Workers and Admin Workers. It will also have methods (processes that can operate on attributes) that can be inherited by Contractors, Project Workers and Admin Workers; for example creation of new name and address, that can be invoked by both creation of Contractor and creation of Employee.

We can look ahead to physical design to see some of the reasons for extending the LDM in the way suggested.

4.6.1 Physical design

We want to be able to implement the design in conventional relational technology – it is not the only option, but we want our specification to be able to support it.

Inheritance

Sub-types should inherit their super-types' methods as well as their attributes. For example, 'Replace worker address' should be the same method whether the Worker whose name is being replaced is an Employee or a Contractor.

If the actual code (rather than just the specification) is to be inherited, then all instances of Worker must be in the same database table. This can be done in two ways:

- the super-type can be implemented in a table separate from the sub-types

- the entire class hierarchy must be implemented in a single table, with mutually-exclusive columns for sub types' attributes.

Clearly, the second option is over-restrictive. Our specification should at least allow for the possibility of the first. Note that distinct specification of super-type and sub-types maps easily on to single-table implementation – the super-type and each sub-type can be a view of the table (updatable since each view is of a single base table).

Encapsulation

Sub-types should be hidden from processes that have no need to distinguish them. For example, when a Workstep is terminated it has different effects on Project Worker, depending on whether the instance is an Employee or a Contractor – but the Workstep process doesn't need to know this. The process that will be invoked for Project Worker should sort out internally what sub-type the instance is, and supply the appropriate effect. This means that we need a single ELH for all the sub-types of a super-type (so that we get a single ECD node, with a selection of effects).

4.6.2 LDM extension

We need to modify the LDM so that we have a general aspect for the super-type, and a specialised aspect encompassing all the sub-types. Note that this is not a formal part of LDM documentation. In practice, it would probably be an expansion of a few local areas of the LDM, simplified after RDA.

There are three guidelines for this modification of the LDM:

- for a simple hierarchy create, at each level, one aspect for the generalised behaviour of the super-type, and one aspect for all the specialised behaviours of the sub-types

- if a sub-type is in more than one hierarchy, create an aspect in each one

- before developing the entity-event model, see if relational data analysis can simplify the LDM.

For example, the type hierarchy of Worker is converted into two aspects for the general aspect of Worker and the employment aspect as in Figure 4.18.

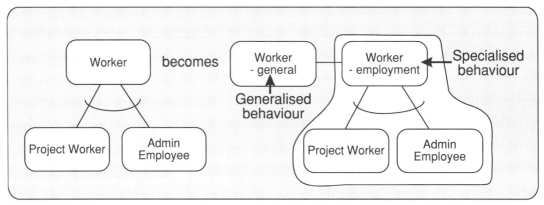

Figure 4.18: Create separate aspects for generalised and specialised behaviours

We could further sub-type Project Worker into Contractor and Project Employee. Project Employee and Admin Employee are also in a hierarchy under Employee; we create other aspects for them there. See Figure 4.19.

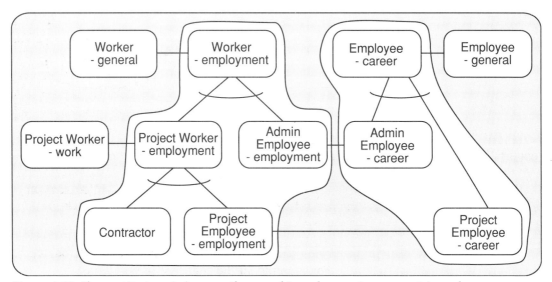

Figure 4.19: If an entity type is in more than one hierarchy, create an aspect in each

Chapter 4
Data modelling extensions

4.6.3 Reconciling the results of RDA with class hierarchies

Class hierarchies result in several LDM components with the same key. For example, in Figure 4.20 there are five types of 'Worker' component that may have attributes, and are identified at the instance level, by Worker Number. (Note that Worker - employment and Project Worker - employment can have no attributes; the common attributes have been 'factored out' into the Worker - general and Project Worker - work aspects.)

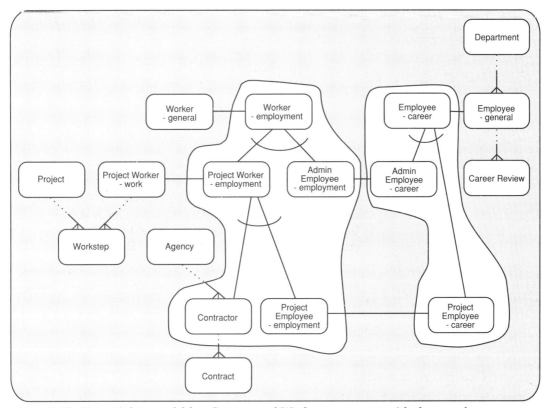

Figure 4.20: 'Aspect' data model has five types of Worker component with the same key

RDA with class hierarchies and aspects is different from the procedure in the *SSADM V4 Reference Manuals*, in two ways:

- RDA sources may use external, user-friendly keys. For example, in the context of a prototyped dialogue, contractor data would be functionally dependent on Contractor Number. But Contractor Number is not a key in the LDM – it is an attribute

55

of the Contractor sub-type, whose (internal) key is Worker Number. We have to take account of this in merging RDA results with the LDM

- there may be multiple relations with the same key. For example, contractor attributes could be assigned to Contractor, Project Worker - general or Worker - general. We have to decide where to allocate each attribute. This is complicated further when there is multiple inheritance – employee attributes could be placed in either of two hierarchies.

The 'aspect' view of the LDM, shown in Figure 4.20, gives us all the possible options for allocation of attributes. Then, when we have completed RDA, if it turns out that some aspects are trivial, we may remove them, simplifying the model before entity-event modelling.

4.6.4 Remove any trivial LDM entities

When we have attributes for each aspect, we have to decide whether any of them can be removed from the LDM. The possibilities are:

- discard super-types with no distinct attributes

- discard sub-types with no distinct attributes

- 'roll up' sub-types with the same user-friendly key into the super-type

- convert class hierarchy into an aggregation.

Discard super-types with no distinct attributes

If a super-type has no attributes of its own, we can discard it and separate the sub-types. In the example, we might have created an 'Employer Organisation' super-type of Agency and Department. After RDA, if there were no attributes other than the (internal) identifier of Employer Organisation, we could discard it.

Discard sub-types with no distinct attributes

If the sub-types have no distinct attributes (given that common attributes should have been 'factored out' into the general aspect), we can discard the sub-types.

Roll up sub-types into the super-type

If sub-types have been created by specialisation, so that they have the same user-friendly key, it may be possible to simplify the LDM by rolling up the sub-types into the super-type.

Convert class hierarchy into an aggregation	If sub-types have been created by generalisation, so that they have different user-friendly keys, it may be possible to create a more robust LDM by defining an aggregate rather than a class hierarchy. The questions to ask are:

- are the sub-types strictly mutually exclusive?
- is the division into sub-types stable over time?

Are the sub-types strictly mutually exclusive?

Only sub-typing which applies to 100 per cent of cases should be shown in the LDM. For example, on a financial application we might at first divide Investments into two sub-types, Stocks with a fixed interest rate and Shares with a profit-sharing dividend, but then discover 'preference shares' which earn both a fixed interest rate and a dividend. Given the loosely defined terminology and mobility of the investment market, it is probably safer to define Investment as an aggregation of interest and dividend paying elements.

In the projects example, could any Employee be simultaneously an Admin Employee and a Project Employee? If so, since they have the same user-friendly key, we could merge Employee and its sub-types into a single entity that has attributes for both sub-types.

Is it possible that some people could be Employees and Contractors at the same time? If so, since Employee and Contractor have different user-friendly keys we would convert the Project Worker class hierarchy into an aggregate, in which Project Worker could have Employee and Contractor aspects simultaneously.

Is the division into sub-types stable over time?

We need to ask of a class hierarchy: Over time can an object instance change from being one sub-type to being another?

If we do not want to keep a history, and the sub-types have the same user-friendly key, we could roll up the sub-types. For example, if an Employee could switch between being an Admin Employee and a Project Employee, we could roll up the sub-types into

Employee, and decide 'Admin' or 'Project' by examining the Employee state indicator.

If we do want to keep history, the original super-type becomes an owner of a set of 'period' entities, and the period is sub-typed.

In the Contractors case study, if one Person can be an employee several times and a contractor several times we could change the model as in Figure 4.21.

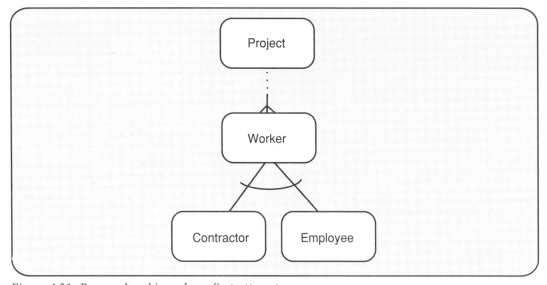

Figure 4.21: Person class hierarchy – first attempt

Contractor case study

In the Contractor case study, suppose that the results of data analysis were:

- Admin Employee and Project Employee have no attributes other than their keys and a code indicating what type of Employee each instance was

- the type code is changeable over time.

The effect is that the class hierarchies can be simplified during data analysis, as shown in Figure 4.22.

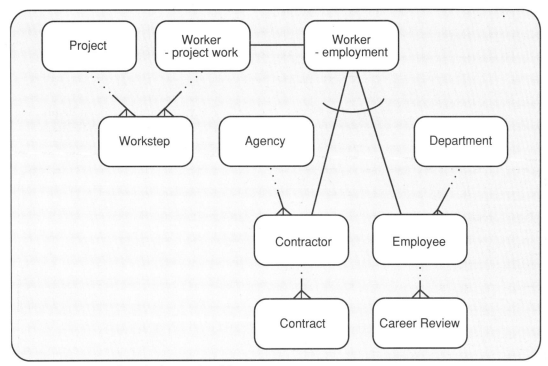

Figure 4.22: Revised Logical Data Model

4.7 Example of achieving reuse

Modelling class hierarchies provides positive analysis benefits. But our aim was to achieve some reuse. How is this achieved?

Consider the representation of workstep assignment to employees and contractors in Figure 4.23

There appears to be duplication of workstep information. Looking at the LDM is not enough to confirm the duplication; we must examine the suppressed details, both data attributes and processing routines. Figure 4.24 shows the data items which are recorded as attributes of the similar-looking entities. (Entity names have been italicised.)

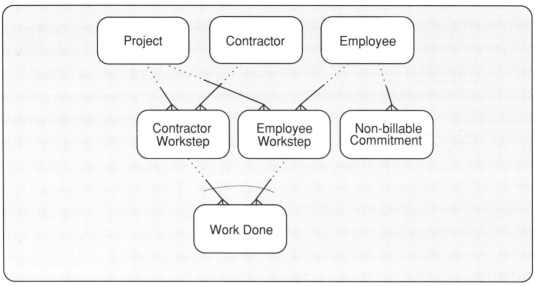

Figure 4.23: Model of workstep assignment with possible duplication of Workstep data specification

Contractor	*Employee*
<u>Contractor Id</u>	<u>Employee Number</u>
Contractor Name	Employee Name
Contractor Agency	Employee Department
	Employee Grade
Contractor Workstep	*Employee Workstep*
<u>Contractor Id</u>	<u>Employee Number</u>
(Project No)	(Project No)
(Workstep No)	(Workstep No)
Start Date	Start Date
Scheduled End Date	Scheduled End Date
Man-day budget	Man-day budget

Figure 4.24: Attributes for similar-looking entities appearing in model of workstep assignment

Both Contractor Workstep and Employee Workstep have two-part compound keys. Apart from the first of their compound key elements, the two entities are exactly the same. What about the behaviours of these two entities and the processing routines defined to implement these behaviours? Let us assume for the moment that they are virtually identical.

This duplication causes unnecessary work for the system builder. For Contractor Workstep and Employee Workstep, two parallel sets of input, update and output processes must be designed and constructed, differing only in the tiniest of ways. To get rid of the duplication, our first thought might be to pass the exclusion arc upwards, combining the two Stock entities as in Figure 4.25.

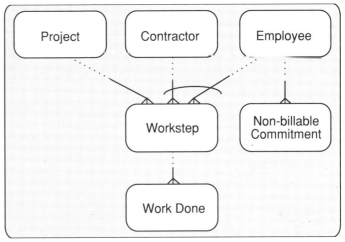

Figure 4.25: Exclusion arc passed upwards

But this model has not solved the problem, since much of the processing will have to be repeated, with slight variations, to account for the two mutually exclusive cases. We want to reshape the model so that the two Workstep entities are truly merged.

It is always possible to turn mutual exclusion between one-to-many relationships into a class hierarchy. For example, Contractor Workstep and Employee Workstep might be regarded as sub-types of Workstep.

The model in Figure 4.26 looks different from Figure 4.25, but it means the same thing. Processing will still be duplicated unnecessarily. The super-type has been introduced at too low a level in the model.

It is better to create a new entity super-type at a higher level, to make a home for attributes or relationships common to both sub-type entities. In the example, Worker is created as a super-type of Contractor and Employee. See Figure 4.27.

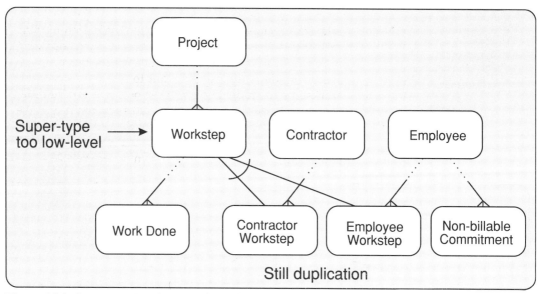

Figure 4.26: Contractor Workstep and Employee Workstep as sub-types of Workstep

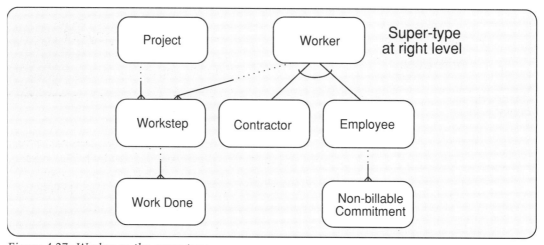

Figure 4.27: Worker as the super-type

All the attributes and relationships common to the sub-types are moved to the super-type entity. Worker takes the attribute Worker Name and the relationship to the detail entity Workstep.

The Workstep entity is now shown as the thing it truly is, the resolution of the many-to-many relationship between Project and Worker.

Like all super-types, it saves a little duplication of data and processing. A bigger benefit comes if we need to introduce a new entity which has all the same characteristics. Suppose we want to allocate some worksteps to students on work experience programmes. For each Student we wish to record their name and the list of worksteps assigned to them. Since a Student has all the characteristics a Worker has, it is easy to plug the new entity into the LDM, as shown in Figure 4.28.

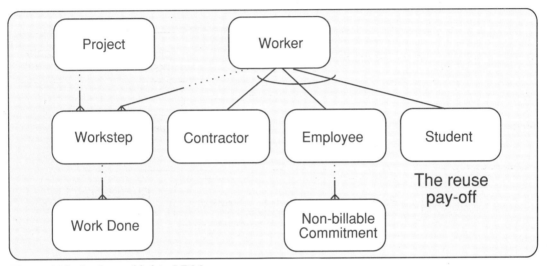

Figure 4.28: Student added to LDM

The main advantage is we do not have to add any data or processing for the Worksteps for Student. The existing database structure and user interface for adding and updating Worksteps will require no revision. This is the reuse benefit we have been hoping for from object-oriented concepts.

4.8 Effect of object behaviour on reuse

In the approach in this chapter, we have identified class hierarchies by examining data and relationships. This is only half of the requirement; object behaviour is as important. For example, in Section 4.7 we focused on super-typing Worker because Contractor Workstep and Employee Workstep had similar attributes and were related to Contractor and Employee – both representing people in the real world. Then we were able to add Student, based on similar reasoning.

We assumed that worksteps behave in the same way – being assigned, scheduled, rescheduled, having work done recorded against them – regardless of what kind of worker they are assigned to. This turns out to be reasonable. There is only one type of real-world object type called workstep, and instances of it always behave in the same way. The only difference is what category of person actually does the work. Whatever the differences between the three kinds of people, they all behave in the same way as far as worksteps are concerned – they are assigned, they report progress by submitting timesheets, and eventually they complete the work and are de-assigned or the workstep is cancelled before they are finished.

Since there are common data and common behaviours, there is a real reuse benefit.

But it does not always work out like this. Suppose we had a similar discussion in a different context, for a company that sells and repairs high-value products such as cameras and personal computers, with instances of products individually identified. The company has warehouses, shops and service depots, and products are stocked at all three types of location. We might finish up with an LDM such as that in Figure 4.29.

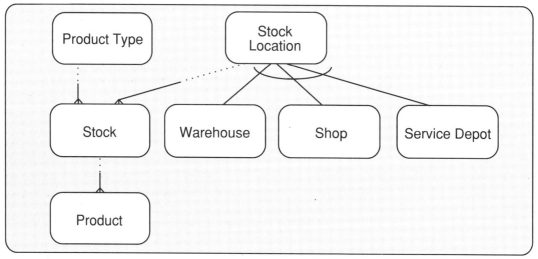

Figure 4.29: Sub-types of Stock Location whose Stocks' behaviours are governed by different business rules

Suppose the business rules were:

- stocks at warehouses are used for supplying shops and service depots; deliveries from suppliers are made only to warehouses

- stock at shop is sold to customers; surplus stock may be returned to a warehouse or sent to another shop

- stock at service depots is used for spares and loans to customers during repairs; it is never returned to warehouse.

The three kinds of stock are different real-world entities. When they are modelled in data they share domains for their attributes – but their attributes from the same domain do not always mean the same thing. 'Warehouse Stock on Hand' means 'stock at this location available for distribution within the organisation', whereas 'Shop Stock on Hand' means 'stock at this location available for immediate sale to customers'.

We also need to consider the behaviour of dependent entities. For example, an individual stock item is constrained as to what can be done with it (transfer, sale, use as temporary replacement etc), depending on what kind of stock it belongs to.

In summary, we can identify possibilities for class hierarchies by examining attributes and relationships, but we should not commit to them in the specification until we have checked out their behaviour, by Entity Life History analysis or some equivalent technique.

4.9 Limits to Reuse

In previous sections we have described how to identify potential super-type entities, and when, and how, to include them in the Logical Data Model. But remember the statement in the introduction to this chapter that event class hierarchies yield more reuse and that we should not expect too much in the way of reuse from entity class hierarchies. This section explains why.

Within the development of a single information system, class hierarchies should be specified wherever they can be found. The duplication of data and processing can be reduced. But many organisations are seeking reuse across information systems. What about the possibilities for reuse between information systems? Below are six reasons why entity class hierarchies are not as reusable between different information systems as we would like:

- class hierarchies are seldom sharable between systems (Section 4.9.1)

- few useful super-types exist (Section 4.9.2)

- new sub-types must match existing super-types (Section 4.9.3)

- a data manager is needed to keep order (Section 4.9.4)

- forethought is needed to get benefits (Section 4.9.5)

- to be useful, a super-type should have a user-friendly key (Section 4.9.6).

4.9.1 Class hierarchies are seldom sharable between systems

To date object-oriented programming languages have been very much concerned with implementing production software such as compilers and screen handling systems, rather than information systems. Part of the reason for the success of object-oriented programming languages in the development of production software is that class hierarchies can be borrowed from one developer by another, then extended for the purposes of the latter.

The world modelled in developing production software is a very restricted one. Varied as operating systems may seem to be, they deal with a limited range of man-made objects *inside a computer*. It is clear that they must all handle similar objects (data item types, records, windows, buttons, devices, blocks, etc). It is easy to imagine that object class hierarchies will turn out to be reusable between such systems.

It is not so clear that business information systems share reusable components to the same degree. The world that is modelled by information systems is a much wider one, of objects in the *business world* and in the *natural world*.

Because information systems deal with substantially different areas of concern, it is less common that a class hierarchy developed for one system will be useful in another.

4.9.2 Few useful super-types exist

Super-types should be modelled wherever they are found within an information system, but in practice we find few super-types are shareable between distinct information systems. Those that are tend to be drawn from the very limited range listed here:

- point in space, or location
- point in time, or date
- person or organisation.

Perhaps the manufacturers of database management systems should be providing these entities for us as standard record data types, rather than data item types.

4.9.3 New sub-types must match existing super-types

Reusability can clearly be achieved where the new entity is to be processed in exactly the same way as an existing super-type entity. But if the data attributes or processing behaviour of the new sub-type (Service Depot) are significantly different from the other sub-types (Shop and Warehouse), then there is less to be gained. Looking for reusable super-types is often more trouble than it is worth.

4.9.4 A data manager is needed to keep order

To achieve the benefits we must build some kind of corporate LDM. This may have to cover several relatively discrete systems, with only small elements in common. Who is in charge of this model? Who changes it to accommodate a new system or user requirement? Who decides what super-types will be useful?

It is dangerous to allow everyone to add new super-types and sub-types into the model. Changes to the LDM must be managed. Probably the responsibility for authorising changes will be given to one person, the data manager. The role of the data manager is vital. Done well, the job will provide opportunities for saving time in system development. Done badly, the job can actually handicap new systems development.

4.9.5 Forethought is needed to get benefits

To speed up future system development, and minimise the redesign necessary, thought may be given now to defining the characteristics of data in the immediately required system, so that it will serve the purposes of future systems. For example, we might make sure that the Stock Location Address attribute is long enough to accommodate foreign country names, in case the business grows to become international.

But such forethought can have harmful side-effects. It can make the immediately required system less user-friendly. It can slow down the development of the immediate system. The future requirements may never materialise.

4.9.6 A user-meaningful super-type should have a user-friendly key

Super-types developed by specialisation are usually meaningful to users (super-types developed by generalisation often are not). But it is hard to make use of a super-type entity if it does not have a user-friendly identifying key. Locations can be identified by a map reference, and possibly height from sea level, and thus cross referenced to a Geographical Information System. Points in time can be identified by a date or a time within a date.

But keeping track of people and organisations by name only is a notoriously difficult problem. People present themselves with many variations of their name, even changing it altogether on marriage. How do we recognise that the same person appears in a medical records system as a doctor, a patient, a pharmacist, a mother, etc? What if one company takes over another?

4.9.7 Reuse from object behaviour and event class hierarchies

It is a mistake to hope for too much reuse from data modelling alone. What we need are further techniques for analysing and specifying the common processes shared between different database programs. It seems there is more reuse to be achieved by developing *event* class hierarchies than by developing *entity* class hierarchies. This is where object-oriented ideas pay the maximum dividend in information systems development and this is what Chapter 5 describes.

4.10 Validating the LDM by entity-event modelling

We should not be surprised if the LDM requires modification during entity-event modelling, indeed we should be surprised if this is not the case. Entity-event modelling should be used to validate and where necessary enhance the LDM. In this way entity-event modelling should help us to reach a 'right' specification. This will not happen where the LDM is 'fixed' and used to constrain entity-event modelling.

5 Process modelling extensions

5.1 Introduction

5.1.1 Event class hierarchies

Data modelling is fundamental to information systems development. However, in most systems, process specification and coding take more effort. In Chapter 4 we used the concept of a hierarchy (super- and sub-types) to support reuse of data. We shall use an analogous concept in processing – event class hierarchies.

Fortunately, class hierarchies of events are both more common and more valuable than entity class hierarchies. Event class hierarchies are more valuable because:

- typically, they yield more reuse than entity class hierarchies (although, in practice, both seem to yield limited amounts of reuse)

- it costs more to specify methods than to specify attributes.

Note that event and entity class hierarchies may be distinct. We can get the benefit of an event class hierarchy without there being any entity class hierarchies in the LDM.

5.1.2 Structure of this chapter

Sections 5.2 and 5.3 extend the process modelling concepts of SSADM in two ways, to support specification of reusable processes:

- addition of operations to Effect Correspondence Diagrams (ECDs), to specify methods for the entity types affected. This allows other options for process specification in addition to Update Process Models (UPMs)

- optimisation of entity state indicator values.

Section 5.4 covers some points of practice to help with the discovery of the 'right' set of events or methods.

Section 5.5 describes how event class hierarchies can be discovered and specified during process modelling in

SSADM, and introduces the concept of the 'super-event' for specification of reusable methods.

Where there is an entity class hierarchy in the LDM, the super-type's methods can be inherited by the sub-types, using the super-event concept.

Section 5.6 describes how the processing for an entity class hierarchy can be modelled as an aggregation of super-type behaviour and sub-type behaviours.

The approach described in Sections 5.4, 5.5 and 5.6 leads to specifications of reusable processes that could be implemented with conventional mechanisms for sub-routine calls.

Encapsulation and message-passing	Section 5.7 describes an approach for encapsulating the methods for an entity type in an 'object class handler'.
	In Section 5.8 the ECD for an event then specifies a network of invocations of object classes. Encapsulation supports easier testing and increases robustness of the system over time.
	Note that encapsulated process specifications require a mechanism for message-passing between object classes; two options are described in Chapter 6.
Enquiry processing	Finally in this chapter, there is a brief discussion of enquiry process modelling in Section 5.9.
5.1.3 Simplified Contractor case study	To examine the development of processing specifications, we use a smaller, slightly different version of the Contractor case study used earlier. In this version, only contractors work on worksteps, and several contractors may work on a single workstep. The assignment of a contractor to a workstep defines a task. Each task must be carried out under single contract. A workstep may be a step of the company's development methodology.
	The LDM, with attributes, is shown in Figure 5.1.

Chapter 5
Process modelling extensions

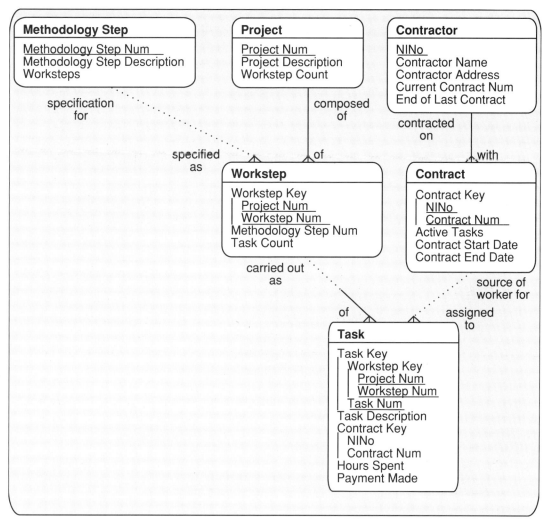

Figure 5.1: Initial LDM for modified Contractor case study

5.2 Methods and operations

5.2.1 What is a method?

A method is the process invoked when an event or enquiry affects an entity. It is represented in SSADM by a node in an Effect Correspondence Diagram (ECD) or Enquiry Access Path (EAP). It could be a single effect, as in Figure 5.2.

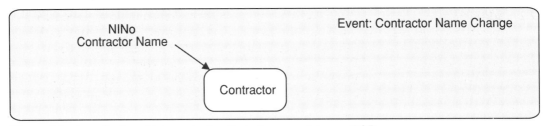

Figure 5.2: A method could be a single effect of an event in an ECD

It might be a selection of effects, as for Contract in Figure 5.3.

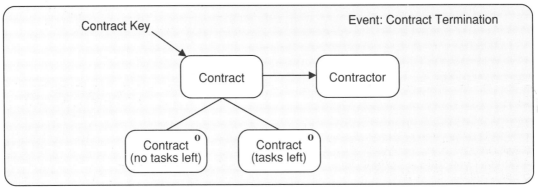

Figure 5.3: A method could be a selection of effects of an event in an ECD

There might be different methods (different effects) for different instances of the same entity type, represented as entity roles. For example, in the earlier version of the case study, two Departments are affected in an Employee Transfer, as illustrated in Figure 5.4.

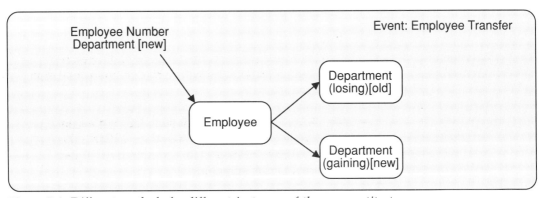

Figure 5.4: Different methods for different instances of the same entity type

Chapter 5
Process modelling extensions

5.2.2 Addition of operations to ECDs

Effects also carry operations. Conventionally, in SSADM V4, operations are not shown on ECDs; they are carried through from Entity Life Histories (ELHs) to Update Process Models (UPMs). But they could be shown on ECDs, and should be if we want to develop process specifications without using UPMs – we shall do this later in this chapter.

Each effect carries two kinds of operations: those that are explicit on the ELH, and those implied by the ELH – get the event input; create, read, write, delete the entity instance; set and check the state variable. The method also needs conditions for any selections. An example of a method with operations and selections is illustrated in Figure 5.5.

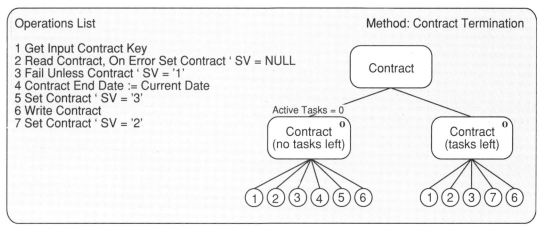

Figure 5.5: A method with common 'front-end' operations in the selection

We need to move the common 'front-end' operations outside the selection, to avoid a recognition problem in the process, as in Figure 5.6.

Note that addition of operations to ECDs does not preclude their being transformed into UPMs – the correspondence rules provided in the *SSADM V4 Reference Manuals* still apply.

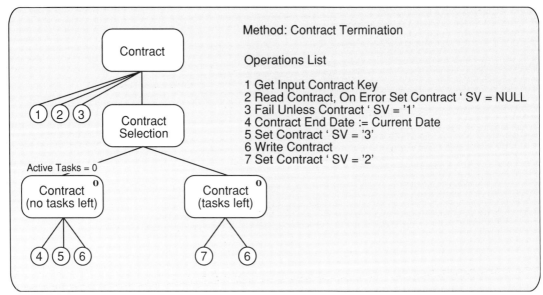

Figure 5.6: A method with common 'front-end' operations moved outside the selection

5.3 Entity state optimisation	In SSADM V4, an entity is given a different state after every event-effect in its ELH. Some states are equivalent, and if equivalent states are optimised to the same state value, reusability can be increased. State values can be optimised with two simple rules, for iterations and selections.
5.3.1 Iterations	The exit state of an iterated component is equivalent to the state before the iteration.
	For example, in Figure 5.7, events A and C may be followed by either event B or event D. Thus the entity's state, in terms of what may be the next acceptable input, is the same after event A or event C.
	In Figures 5.7 to 5.9 optimised state values are shown in bold below the distinct state values.
5.3.2 Selections	The exit states for each option of a selection are equivalent – except when one option ends in an iteration.
	For example, in Figure 5.8, event Z can be preceded by either event W or event Y.

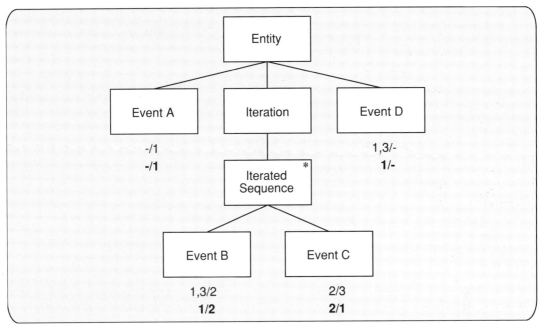

Figure 5.7: The exit state of an iterated ELH component is equivalent to the state before the iteration

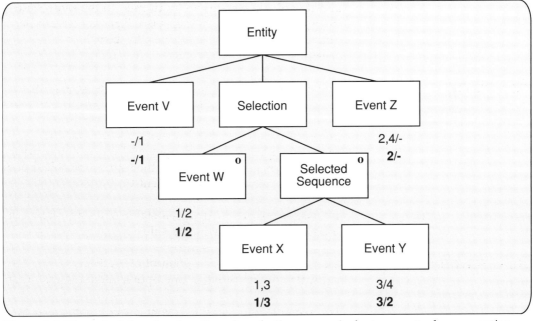

Figure 5.8: The exit states of each option of a selection are equivalent - except when one option ends in an iteration

The state optimisation rules can be used on ELHs containing selections and iterations, except where one or more options of a selection end in an iteration.

For example, in Figure 5.9, we can use the iteration rule to optimise the exit state value for event W to '1'. But we cannot use the selection rule to optimise the exit state for event Y to '1'. If we did, state checking would permit an event of type W to be preceded by an event of type Y.

To prevent this kind of problem, there is an exception to the selection rule: any option of a selection that ends in an iteration must have an exit state different from those of other options of the selection.

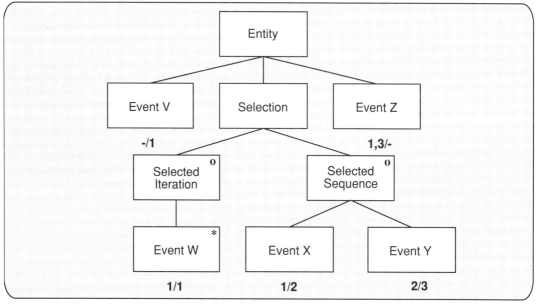

Figure 5.9: There is an exception to the selection rule where a selection ends in an iteration

5.4 Discovering the 'right' set of events or methods

SSADM identifies methods by taking an event-oriented approach to requirements capture and knowledge acquisition. There are two parts:

- first, identification of the basic methods as in SSADM V4 by:

  – looking for events that affect each entity type

- seeing if there are multiple effects on an entity type

- seeing if different effects apply to different entity roles.

Points of practice are summarised in this section.

- second, identification of reusable methods by:

  - creating 'super-events', as an extension of the SSADM V4 approach for building ELHs.

Super-events are described in Section 5.5.

5.4.1 Points of practice for finding and naming basic methods

Initially assume that each event fires a unique method in an entity. Thus the initial list of methods can be taken directly from the initial list of events.

This one-to-one correspondence remains true for the majority of events and methods. However, ELH analysis normally reveals a more subtle relationship between events and methods.

Name events after the triggers of update effects on entities

An ELH should include the events which trigger updating effects on an entity. Updating effects can be classified into five (not mutually exclusive) kinds:

- entity birth effects

- entity death effects

- relationship change effects

- relationship attribute change effects

- entity state change effects.

A single event can have one or more of these five effects upon an entity. We identify events by looking for the causes of these effects. We do this by asking what events trigger operations upon the 'aspects' of an entity.

In addition to the update effects of events, an ELH may include the validation effects and enquiry effects of events which update other entities.

Do not name methods after implementation operations	Method names like Store, Modify and Delete should be avoided wherever possible because these are things which are done to the physical representation of an object rather than to the conceptual object. Method names need to be taken from the same universe of discourse as the objects, not the universe of discourse of the implementation environment.
Initially omit trivial attribute changes	Many information systems store a large number of simple non-key attributes, which can be replaced individually without any knock-on effect elsewhere in the system. These attributes might lead us to identify numerous simple events, one for each attribute, where each event merely replaces the current value of an attribute with a new value.

ELH analysis is often more manageable if the simple attribute-changing events are postponed to a second pass. To begin with, we concentrate on the events that create and kill off entities, make and break relationships and change entity states.

Alternatively, it may be possible to simplify the entity-event model by lumping together all the trivial attribute change events for one entity type, as a general Replace Entity Details event. This works if it can be assumed that unchanged attributes can be re-input along with the changed ones without any side-effect on the system, as is possible with some application generators.

Do not make an event too big or too small	An event is a small parcel of system behaviour. On entering the system, an event will trigger an update process which will move the system from one consistent state to another and *cannot* be partially executed.

An event is atomic in the sense that it cannot be divided. By definition, if the process which applies the event to the system is only partially executed, then the system will be left in an inconsistent state.

Divide composite events into smaller distinct events

Those new to event identification often define as events things which are much larger than events and which are really more like user functions. Do not define a series of things which could happen independently as a single event.

For example, consider an input document which notifies us of the birth of several entities of the same type. It is better to regard the document as being composed of several distinct birth events, than to define an event for the whole document which will be the birth of an arbitrary number of entities.

Do not split one event into distinct events

Whatever must succeed or fail as a whole cannot be split into two or more events. For example, if it is a rule that an applicant must belong to an office, then a transfer of an applicant between two offices should not be split into distinct 'dispatch' and 'receipt' events.

Give each entity its own birth event

In general, each entity type has its own distinct type of birth event. But a dilemma often arises where the user wishes to create a set of detail entities at the same time as a master entity. Is this one event, or a series of distinct events? Here are two rules:

- if a master entity *can* legitimately exist without detail entities, then the birth event of the master entity should not create detail entities and the detail entities *must* have their own birth event

- if a master entity *cannot* legitimately exist without detail entities, then either:

  - the birth event of the master entity should create one or more detail entities as well and the detail entities *may* have their own birth event

  - or the birth of the master is the birth of detail (first).

5.5 Identification and analysis of reusable methods

5.5.1 Reusable methods and super-events

Using basic ELH analysis as described in the *SSADM V4 Reference Manuals*, the same processing routine (or method – in fact, an ELH documents method invocations) may be defined more than once, as triggered by different events. Recognising this reusable method as being caused by a **super-event** leads to some simplification of ELHs, ECDs and UPMs.

Super-events enable common code to be detected in the ELHs, and specified as routines invoked by several UPMs (these routines being recorded as ECDs). Whenever a new event is added under a super-event in an ELH, it can use the common routine already tested and implemented.

5.5.2 Identifying super-events

A super-event occurs where:

- two or more event-effects share the same operation set

and

- end in the same entity state (after state optimisation as described in Section 5.2.2).

A super-event is usually recognised in the ELH as a selection component above effects for different events, as illustrated by '§Workstep Definition' and '§Workstep Termination' in Figure 5.10. Note that the '§' prefix is not a mandatory notation, merely a means of highlighting super-events within this volume.

Chapter 5
Process modelling extensions

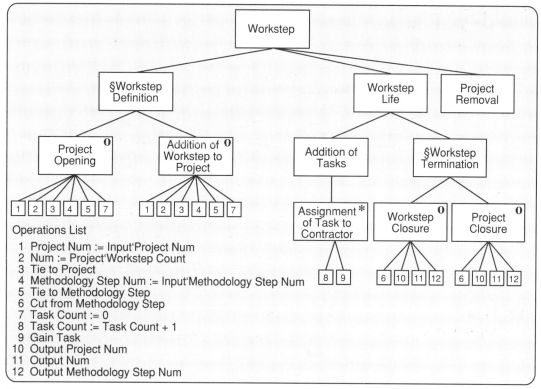

Figure 5.10: *Super-events as selection components*

§Workstep Definition is a super-event which appears only once in the ELH. There will be an ECD for each of the events that invoke the super-event, 'Project Opening' and 'Addition of Workstep to Project'. See Figure 5.11. The Workstep effect in each has only one operation: 'Invoke §Workstep Definition'.

Note that in 'Addition of Workstep to Project', the entry point is Project, since Project Number is given as input, and the next Workstep Number is assigned using Workstep Count in Project.

83

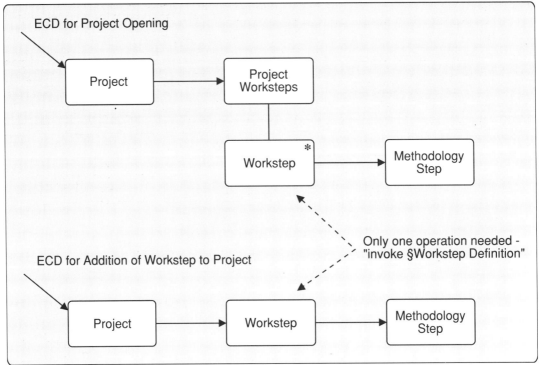

Figure 5.11: ECDs from which §Workstep Definition is invoked

There will be a separate ECD for the super-event as in Figure 5.12, to which operations have been added to indicate the reused processing.

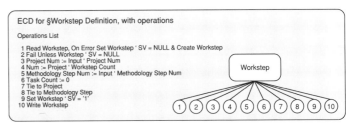

Figure 5.12: ECD for §Workstep Definition with operations

The reuse benefit obtained here is small. A more significant benefit can be found where the super-event appears more than once, in one or more ELHs.

5.5.3 Super-event used across more than one ELH

The events that invoke §Workstep Definition and §Workstep Termination also affect Methodology Step. A simple ELH is shown in Figure 5.13.

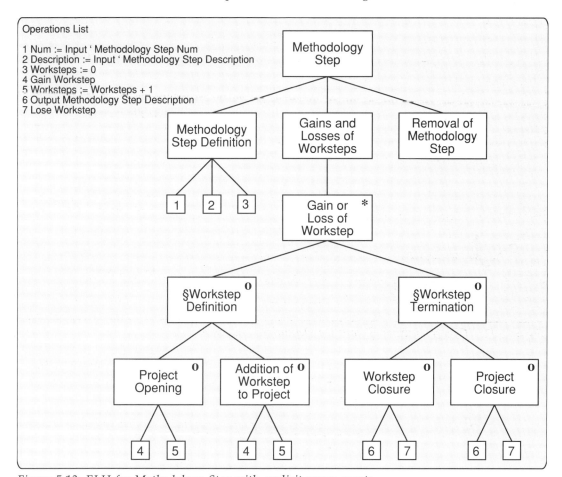

Figure 5.13: ELH for Methodology Step with explicit super-events

But if we recognise the super-events when creating the Workstep ELH, and check that they apply also to Methodology Step, we need include only the super-events in Methodology Step's ELH. This simplifies the ELH, as in Figure 5.14, and increases reuse.

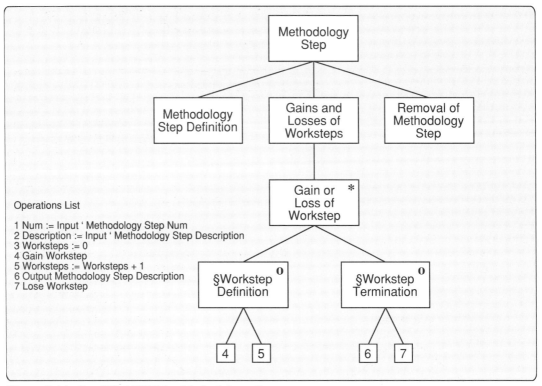

Figure 5.14: *Simplified ELH for Methodology Step*

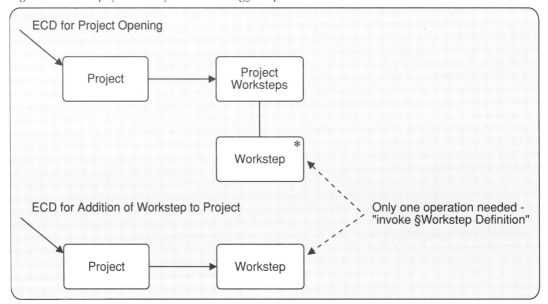

Figure 5.15: *ECDs for base events*

The effect on Methodology Step is moved out of the ECDs for the base events (see Figure 5.15) and into the ECD for §Workstep Definition (see Figure 5.16).

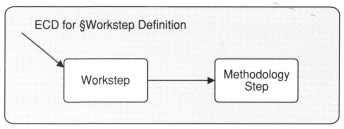

Figure 5.16: ECD for §Workstep Definition

5.5.4 Documentation of event class hierarchies

As reusable components are defined, we need to document the hierarchies of events and super-events. For example, as shown in Figure 5.17.

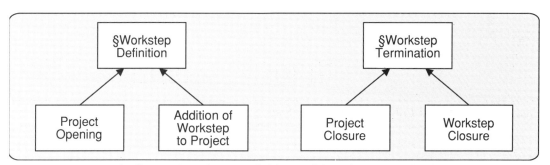

Figure 5.17: Events leading to §Workstep Definition and §Workstep Termination

Super-events can call other super-events. For example, §Workstep Termination and Completion of Task define a super-event in Contract. See Figure 5.18.

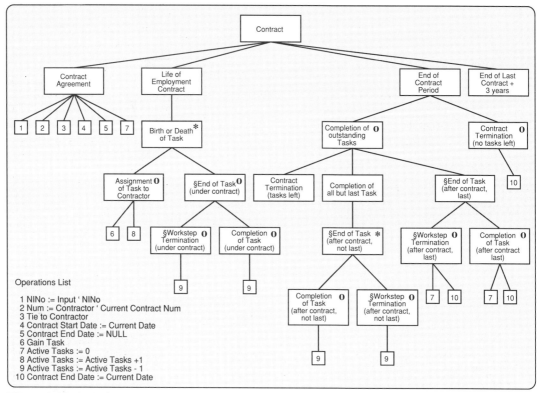

Figure 5.18: §Workstep Termination and Completion of Task define §End of Task

The higher-level invocations have to be recorded in the event hierarchies as in Figure 5.19. We also need to record where in the ECD, ie on which entity type, the invocation of the super-event will be made.

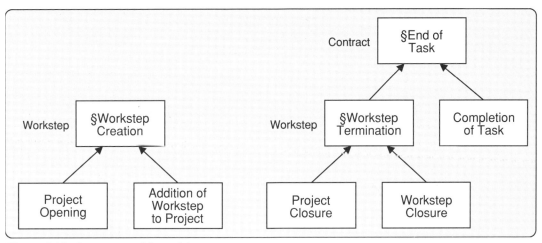

Figure 5.19: Super-event hierarchies

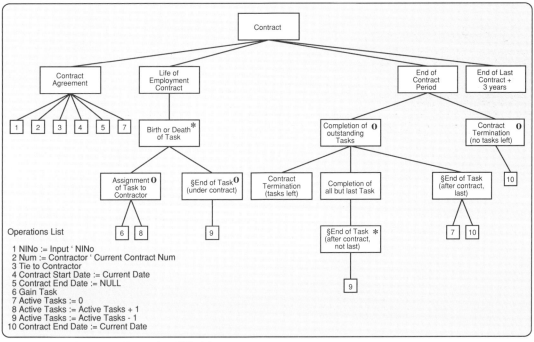

Figure 5.20: Contract ELH simplified by super-events

If we establish, in a super-event hierarchy, the links between base events and super-events, we can omit the base events from the ELHs. Figure 5.20 shows a simplified ELH for Contract.

5.5.5 Effect of omitting base events from ELHs

If we omit base events from ELHs, we have to take account of the links between events and super-events when constructing ECDs.

For example, Project Opening, Addition of Workstep to Project, Workstep Closure and Project Closure do not appear in the simplified ELH of Workstep (see Figure 5.21), although they will appear in Workstep's column of the Event-Entity Matrix.

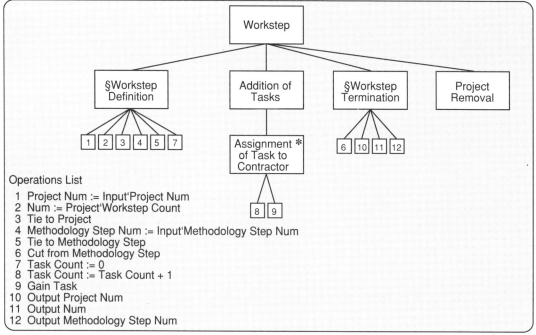

Figure 5.21: Simplified ELH for Workstep

We have to use the references in the super-event hierarchy (and the Event-Entity Matrix) to include an effect for Workstep in the ECDs of each of these events. This effect contains the operation to invoke the common process defined by the super-event. Figure 5.22 shows the ECDs for Project Closure.

Note that it is not necessary to replicate the super-event ECDs in the ECD for each base event. Figure 5.22 simply illustrates the relationships between ECDs.

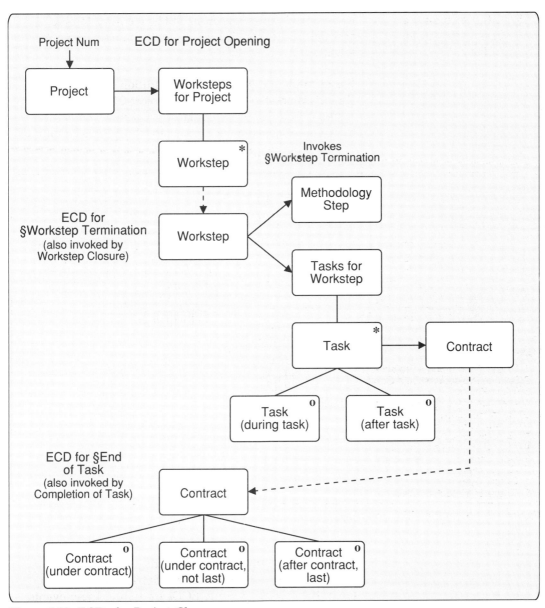

Figure 5.22: ECDs for Project Closure

5.5.6 Process models for super-events

Update Process Models (UPMs) provide a useful processing specification for some types of implementation; for example, host/SQL for database servers.

To accommodate super-events, the SSADM V4 guidance for transforming ECDs into UPMs requires two minor extensions:

- 'invoke' and 'return' operations are needed; note that failures are also 'return' statements

- if an entity, such as Workstep or Contract in Figure 5.22, has effects in both an invoking ECD and an invoked ECD, its 'read' operations are placed in the *invoking* UPM. Note that other conventions can be made to work, but require more extensive guidance.

Figure 5.23 shows the UPMs for Project Closure, §Workstep Termination and §End of Task, produced with the extended UPM guidance. For some implementation environments we shall have to specify data and/or message areas for invoke and return operations; we can leave this until later.

In accord with the guidance given above, the read operations for Workstep are in the UPM for Project Closure. This means that §Workstep Termination can be invoked iteratively from Project Closure and singly from Workstep Closure. If we had used a convention that placed the read operation in §Workstep Termination, the UPMs would have had to be more elaborate to control the iteration and handle the failure conditions.

In some cases, the ECD and UPM for the base event are minimal. For example, the UPM for Workstep Closure only has operations to get the input, read the Workstep and invoke §Workstep Termination.

Chapter 5
Process modelling extensions

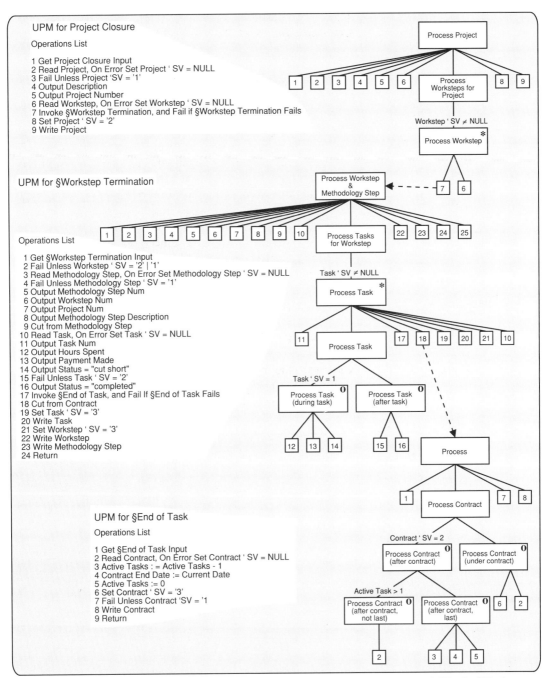

Figure 5.23: Extended UPMs for Project Closure, §Workstep Termination and §End of Task

5.5.7 Limitations of super-events

Adding super-events to ELH analysis has some drawbacks.

More complex meta-model
Super-events complicate the meta-model of the methodology. It is no longer possible to see everything that one event does on a single ECD, since it may invoke super-events, which themselves call super-events, and so on.

Indirect references on ELHs
If we use super-event names on ELHs and omit the base events, we lose the direct references in the ELH from entity to events (although these references can be preserved in the Event-Entity Matrix). We have to use the event class hierarchies (as illustrated in Figures 5.17 and 5.19) to see the full effects of events on entities.

Super-events give us more to think about
We have to look for events with common effects, then decide whether to create a super-event or not. Two rules of thumb are:

- if two events have the same effect on an entity, but this common effect does not or cannot appear anywhere else as a super-event, it may be worthwhile creating a super-event just for the one effect (but it may be worth considering a 'common operation set', described under the next sub-heading

- if two events have the same effect on an entity, but arrive at that entity from different directions, do not bother to create a super-event. The two events are likely to go on to diverge, having different effects on different entities. It may be worth considering a 'common operation set', described next.

    This often happens as a result of death events' being copied down from two different masters into the detail entity that links them.

Common operation set
Super-events do not help us identify what might be called 'common operation sets', that is groups of operations that are repeated under different effects. In

practice, this is probably an advantage, since searching for reuse between the level of an event effect and the operations allocated to that effect is probably more trouble than it is worth. However, common calculation routines may be specified as single operations.

It may be worthwhile creating a common operation set for the 'one-off' candidate super-events recognised in the two rules of thumb above. The base events are placed in the ELH, with one operation: 'invoke *common-operation-set-name*'.

5.6 Entity class hierarchies and inheritance

So far we have looked at re-usability by using event class hierarchies to define super-events. We can use the same concepts to model inheritance of methods in data hierarchies.

We need to extend the example a little. Suppose that a contractor can work on an individual contract or under an agency contract covering several contractors. Then, from the work assignment view, a 'Contract' is either an Individual Contract or an Agency Contract Line. The revised Logical Data Model is shown in Figure 5.24.

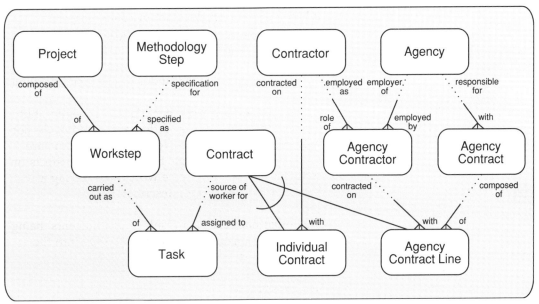

Figure 5.24: Extended LDM

As described in Chapter 4, the approach we shall take is to define two aspects – one for the generalised behaviour and one for the specialised behaviours. See Figure 5.25.

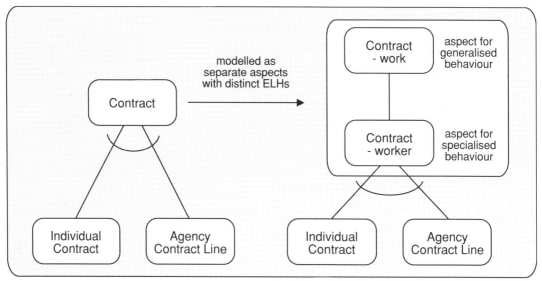

Figure 5.25: Separation of class hierarchy into aspects

The modified LDM is shown in Figure 5.26.

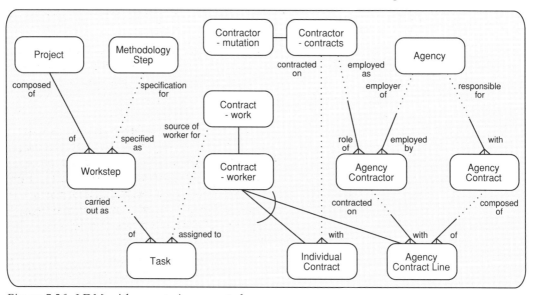

Figure 5.26: LDM with aspects incorporated

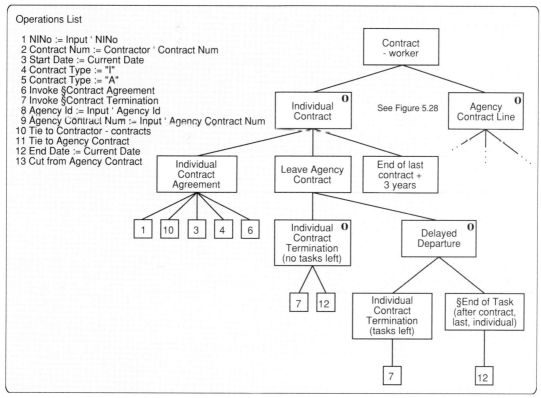

Figure 5.27: Part of ELH for specialised behaviours of Contract

Note that there is no shared behaviour (and no common attributes) in Contract - worker. The ELH is a high-level selection. All shared attributes and behaviour are in Contract - work.

We model Contract - worker in a single ELH to isolate Contract - work from the sub-types. In ECDs (as in the examples which follow), Contract - work should correspond one-to-one with Contract - worker. If we add further sub-types to Contract - worker (or remove sub-types from it) the changes should be confined to effects on Contract - worker. There should be no need for changes to effects on Contract - work.

Note also that aspects are not concerned only with modelling hierarchies. Contractor has parallel lives, modelled in two aspects. Contractor - contracts models a constraint that at any time a contractor may either have

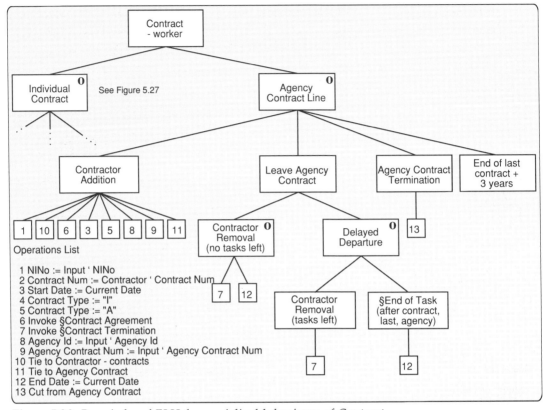

Figure 5.28: Remainder of ELH for specialised behaviours of Contract

an individual contract, or be an agency contractor (records of both are kept for history, so the relationships are not mutually exclusive on the LDM).

But we require to be able to change a contractor's personal details at any time, regardless of his agency or contract status. We model this as a parallel life in Contractor - mutations.

ELH for distinct behaviours

We develop one ELH for all the specialised lives, a high-level selection with one option for each sub-type. For readability, it is shown in two parts in Figures 5.27 and 5.28. Note that we have included operations to invoke super-events in the general life, although (if we had followed the convention we used earlier) they could have been left until ECDs were produced.

ELH for common behaviour

We develop a separate ELH for the general life. Note that structurally it is similar to the ELH of Contract before introducing the hierarchy. See Figure 5.29.

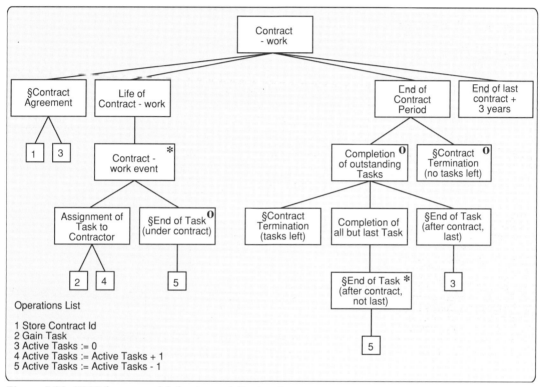

Figure 5.29: ELH for general behaviour of Contract

The differences are that some of the operations have moved into the specialised lives, and that Contract Agreement and Contract Termination are now super-events invoked by events in the specialised lives. Thus, modelling of the general life as a separate aspect allows events in the specialised lives to inherit common processing in the general life, by invocation of the corresponding super-event.

For example, §Contract Termination is invoked by either Individual Contract Termination or Contractor Removal, as illustrated in Figure 5.30. Note that whether there are tasks left or not is determined within §Contract Termination. In the process models developed from the ECDs in Figure 5.30, §Contract Termination has to be invoked before the selection in the invoking process can be dealt with.

As mentioned in Chapter 4, modelling of general and specialised lives as separate aspects does not commit us to implementing them in separate tables.

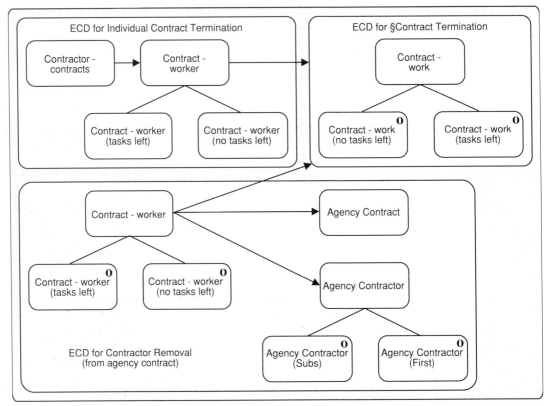

Figure 5.30: *Inheritance of behaviour by invocation of a super-event*

Inheritance may work by the invocation of a specialised life from a general life. For example, Figure 5.31 shows the ECD for §End of Task, invoked from the 'work' part of the LDM.

If §End of Task removes the last Task for a Contract, after a decision to close the Contract, it sets the Contract's work and worker aspects to 'history'.

A single ECD for Contract - worker is invoked. The process derived from this ECD will select the identified instance of Contract, determine which sub-type it is and apply the appropriate effect. This mechanism supports addition of further sub-types of Contract - worker without any change to processing for Contract - work.

Chapter 5
Process modelling extensions

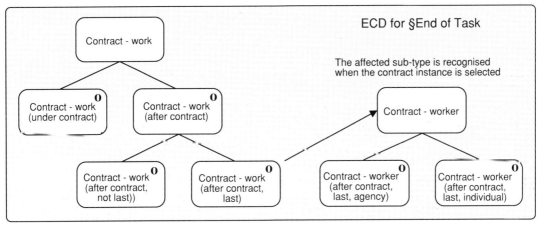

Figure 5.31: Invocation of specialised Contract behaviour from general

5.7 Encapsulation of methods for an entity type

We could encapsulate all the methods for an entity aspect in a single object class component. The object class would be responsible for accepting an input message, selecting the appropriate instance of the entity aspect, invoking the appropriate method and responding to the invoking environment, as illustrated in Figure 5.32.

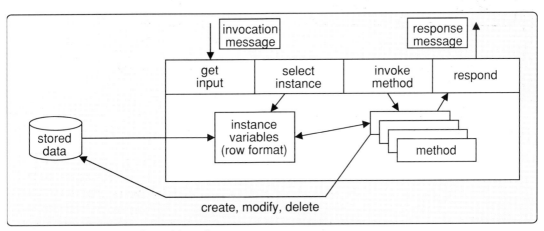

Figure 5.32: Operation of an object class

5.7.1 Processing model for method bodies

Each entity-aspect-role on an ECD defines a method. We saw in Section 5.2 that we specify the method bodies (the processing carried out when methods are invoked) by adding operations to the ECD.

To produce the processing model for a method body, we simply omit the operation to get the input.

101

The process model for the method body in Figure 5.6 is shown in Figure 5.33.

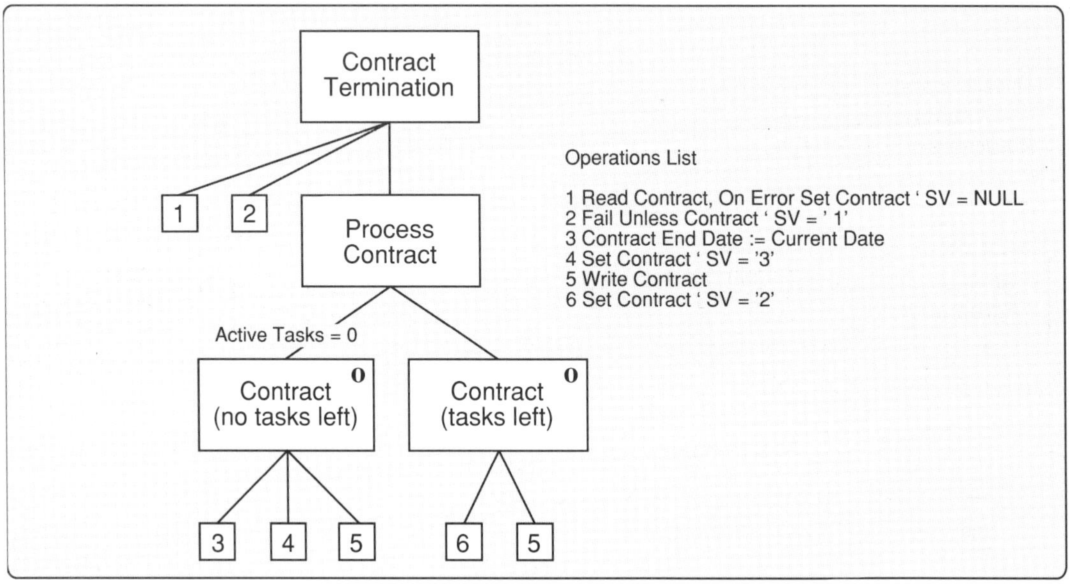

```
                    Contract
                   Termination
                        |
                 -------+-------
                /       |
              [1]  [2]  |              Operations List
                    Process
                    Contract           1 Read Contract, On Error Set Contract ' SV = NULL
                        |              2 Fail Unless Contract ' SV = ' 1'
         Active Tasks = 0              3 Contract End Date := Current Date
                        |              4 Set Contract ' SV = '3'
              ----------+----------    5 Write Contract
             /                     \   6 Set Contract ' SV = '2'
       Contract  O          Contract  O
      (no tasks left)      (tasks left)
         /  |  \              /  \
       [3] [4] [5]          [6] [5]
```

Figure 5.33: Process model for method body for message 'Contract Termination' on object class 'Contract'

5.7.2 Procedure for selection of methods by object class

As well as the method bodies, we need a procedure for each object class (entity aspect) to invoke its methods. After method bodies have been extracted from an ELH, what is left collapses into a selection of methods controlled by input message type (based on event type).

If we add operations to accept the input message and to send the response created by the invoked method body, we have the basis of an object class handler. For example, see Figure 5.34.

5.7.3 Integrating method bodies

We can plug method bodies into their object class handler, as illustrated in Figure 5.35. A CASE tool could derive this structure automatically from the ELH.

5.8 Effect correspondences

In the process models we have developed so far, we have translated effect correspondences into direct association between effects, in one of two ways:

- merging the operations for corresponding effects into a single process node (as in the UPMs in the *SSADM V4 Reference Manuals*)

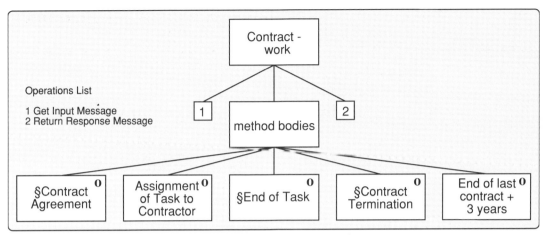

Figure 5.34: Basis of an object class handler for Contract - work

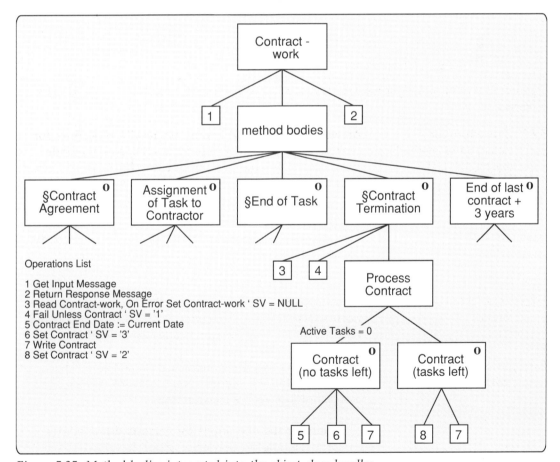

Figure 5.35: Method bodies integrated into the object class handler

- direct invocation of common processing identified via super-events.

If we encapsulate the processing for an object class (entity aspect), as described in Section 5.7, correspondences have to be translated into invocations of object classes. This means that instead of two corresponding effects being merged (or one directly invoking the other), one has to invoke the *object class* for the other. The invoked object class has to select the appropriate method (effect process), as in Figure 5.36.

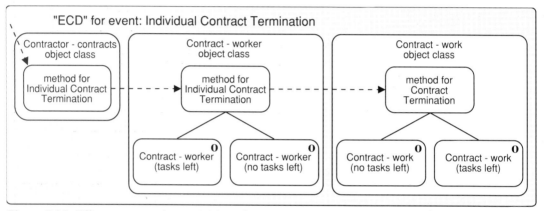

Figure 5.36: Effect correspondence – object <u>class</u> is invoked and knows from event type which method to select

This view of correspondences has three impacts on process modelling. Two we shall deal with here – super-events and iterations. The third, how object classes can be invoked, and messages passed between them, is the subject of Chapter 6.

5.8.1 Effect on process modelling of super-events

Normally, an object class knows which method to use – it corresponds to the event type. But if the method to be used is the start of a super-event (the entry point of the super-event ECD) the same method can be invoked for different event types. For example, in Figure 5.36, the method used in Contract - work is §Contract Termination, which can also be invoked for the Contract Removal event.

In the object class, the input event types that require a method to be invoked have to be specified as the selection criteria for that method, as in Figure 5.37.

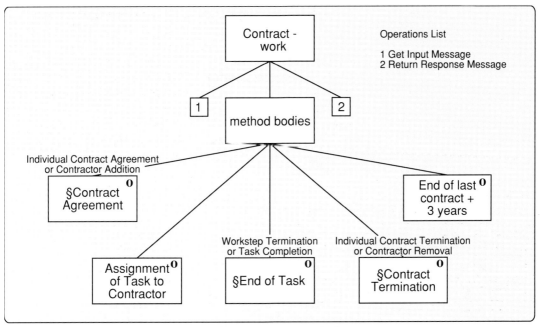

Figure 5.37: Multiple selection criteria for super-event methods

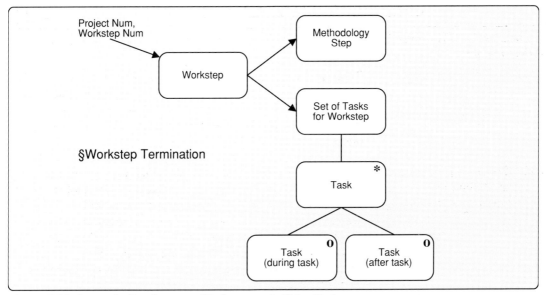

Figure 5.38: Iterated effect between Workstep and all its Tasks

5.8.2 Effect on process modelling of iterated methods

We also have to handle one-to-many relationships, shown as iterated correspondences on ECDs, for example between Workstep and all its Tasks in Figure 5.38.

The following rule of thumb seems reasonable:

- if one input message identifies a set of instances (ie is a non-unique attribute), define a class method for the 'detail' object class. The 'detail' class receives a single input message and applies the appropriate method to all instances affected by the event.

  This is the more common type of iterated correspondence

- if an individual message has to be constructed for each 'detail' instance, handle the set in the 'master' object's method (or in the event manager – described in Chapter 6 – depending on the message-passing protocol selected). The 'detail' class receives a separate input message for each affected instance.

In the example in Figure 5.38, Workstep does not know the individual Task identifiers – all Tasks for the Workstep are affected, so the iteration is handled in Task, as illustrated in Figure 5.39. In a relational implementation, this would probably be a SELECT of all tasks for the workstep, then a cursor to step through the selected tasks and build the output.

Note that operations dealing with the invocation of another object class (place-marked by 'operation' 8), are not defined. Their form will depend on the approach selected for message-passing, as described in Chapter 6.

5.9 Enquiry modelling

There are, broadly, two kinds of enquiries to be considered – those designed into the system and ad-hoc enquiries after the system has been developed.

5.9.1 Designed-in enquiries

For designed-in enquiries, if we add operations to Enquiry Access Paths (EAPs) we can use an approach similar to that described in this chapter for event processing.

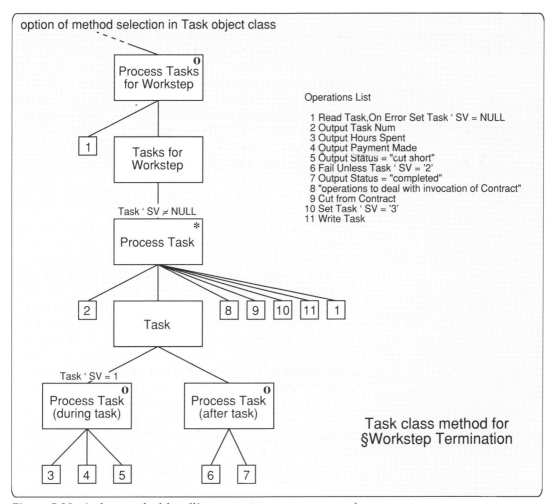

Figure 5.39: A class method handling a one-to-many correspondence

Designed-in enquiries may be freestanding (ie they can be invoked ad-hoc by users) or internal, invoked by event processes or other enquiries. Some internal enquiries are re-usable, and can be specified in the same way as the common processes defined by super-events.

5.9.2 Ad-hoc enquiries

For many systems implemented in relational technology, users require an SQL interface to support enquiries that are not designed into the system – the enquiry specifications are ad-hoc.

A problem with moving to an implementation in which object classes were encapsulated is that it would, strictly, prevent people from writing or generating ad-hoc SQL queries. SQL assumes access to tables of instance data, and encapsulation requires that instance data be accessed only through the object classes.

Solutions to this problem are on the way. An OO extension to the SQL standard is currently under development. Relational technology vendors are developing approaches based on working drafts of the standard.

An interim solution may be to expose the physical design to an SQL or forms interface, for read-only access. Ad-hoc SQL queries will not be as robust as designed-in enquiries; they may have to be rewritten if the database design is altered. To minimise the impact, SQL queries should generally be defined on views rather than base tables.

6 Method invocation and message-passing

6.1 Introduction

If we require methods to be encapsulated, as described in Chapter 5, we must have a mechanism for passing messages between object classes. Implementation can be based on the invocation mechanisms of typical programming languages.

This chapter describes:

- the derivation, from the ECD, of the message-passing requirement and the types of message to be specified (Section 6.2)

- two approaches for specifying message-passing (Section 6.3)
  - a network structure, based explicitly on the ECD
  - a simplified structure, where an event manager invokes each object class directly

- the implications of using network invocation (Section 6.4)

- the implications of using direct invocation (Section 6.5)

- a summary of the strengths and weaknesses of the two approaches (Section 6.6).

Wherever there is a requirement to distribute processing, whether for business process modelling or geographical reasons, there will be a need to address the method invocation and message-passing issues dealt with in this chapter regardless of the final implementation environment.

6.2 Deriving message-passing requirements

Message-passing requirements are represented by the correspondence arrows on Effect Correspondence Diagrams (ECDs), as illustrated in Figure 6.1.

Figure 6.1 shows the ECD for the super-event §Workstep Termination, which invokes a further super-event, §End of Task. Each correspondence arrow indicates a requirement for message-passing between the two object classes.

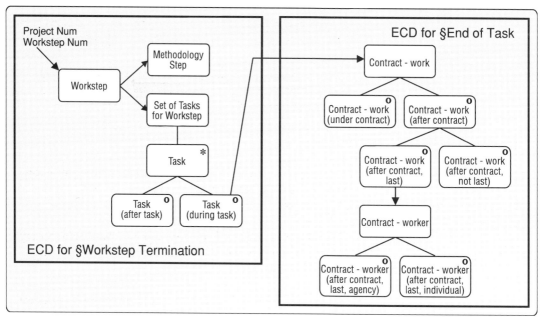

Figure 6.1: Correspondence arrows on ECDs indicate message-passing

Simplified ECDs   If the message bodies are hidden (because they will be encapsulated in their object classes) the ECD collapses into a simpler diagram, as in Figure 6.2, showing the requirements for message-passing between object classes.

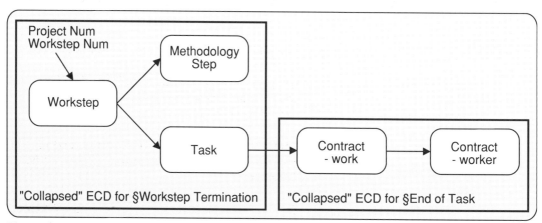

Figure 6.2: ECD with method bodies suppressed

6.2.1 Types of message   For each correspondence arrow there are three kinds of message to be specified, as illustrated in Figure 6.3.

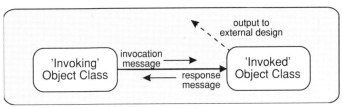

Figure 6.3: Message types

There is an invocation message, sent from the invoking object class, and two kinds of message from the invoked object class:

- response returned to the invoking class, which may be simply 'success' or 'failure', or may also carry some data

- output to the External Design, not required in every case.

Hidden messages

One of the reasons that message handling sometimes looks complicated is that, in most implementations, the response and the external output are packaged together and routed as a single message.

Invocation messages

When an object class is invoked the invocation message needs to tell it:

- the message type (the event type or super-event)

- the event or super-event instance identifier (for selective locking, described in Internal Design in Chapter 7)

- the data items (usually a key or a partial key) to identify the instance or set of instances affected by the message

- the data items needed by the method body. When specifying the message, these are identified by inspection of the operations in the method body.

Response messages

A response message has to contain success/failure status and response data items (where needed). Whether it also needs to identify the event type and instance will

		depend on the implementation approach for message-passing.
	External output message	An external output message will contain the required output data items. Whether the external output message also needs to identify the event type and instance will depend on the implementation approach for message-passing.
6.3	**Two approaches for implementation**	Implementation approaches for message-passing can be broadly classified under two headings:

- network invocation
- direct invocation.

	Event manager object	Both approaches use an 'event manager', an ephemeral object that models the event (ie the commit unit). The event manager is part of the **Conceptual Model** in the 3-schema Specification Architecture. It does not change if the mapping of events and enquiries into functions changes, or if the I/O technology for the user interface changes.
6.3.1	Network invocation	Network invocation models the ECD more-or-less directly, and passes messages from one entity to the next, as in Figure 6.4.

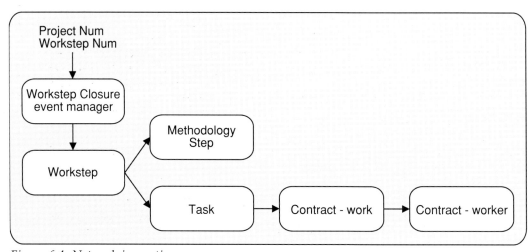

Figure 6.4: Network invocation

Note that the same message-passing convention should be used, whether messages are between the event manager and the first entity-based object class or between entity-based object classes. The invoked object class should behave consistently, whatever object class invokes it.

In the example in Figure 6.4, Workstep is invoked by the event manager for Workstep Closure. But the ECD is for §Workstep Termination, which is also invoked by Project in the ECD for Project Closure.

Workstep should respond in the same way to a §Workstep Termination message whether the invoking object class was Project or Event Manager for Workstep Closure. An invoked object class should not need to know what class sent the invocation message.

6.3.2 Direct invocation

Direct invocation uses an event manager to construct individual messages and invoke object types directly, as illustrated in Figure 6.5.

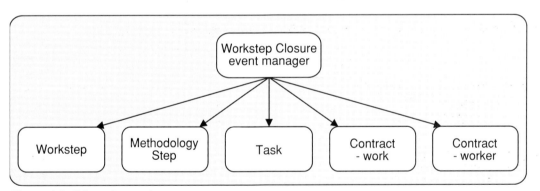

Figure 6.5: Direct invocation

6.4 Implications of using network invocation

Network invocation is easily derived from the ECD. The main design issues are:

- handling of 'passed-through' data
- specification of message structure and content.

6.4.1 'Passed-through' data

Consider the ECD for Assignment of Task to Contract, shown in Figure 6.6 with corresponding arrows annotated with the data items in invocation messages.

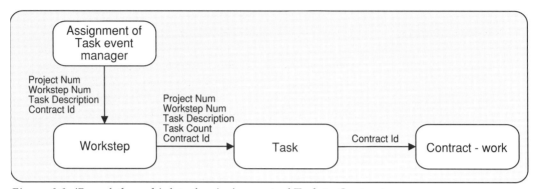

Figure 6.6: 'Passed-through' data for Assignment of Task to Contract

Workstep is needed first because its Task Count provides the specific part of the (user-friendly) key of Task.

As well as its key (Project Num, Workstep Num, Task Num) Task needs a Task Description and Contract Num. These items are passed through Workstep – ie Workstep does not use them, but needs them to construct the Task invocation message.

This means that if there were changes in specification of the data required by Task (eg say a man-day budget were added), the method for Workstep would have to be changed to redefine the invocation message. This should not be necessary.

6.4.2 'Sealed packet' protocol

The solution is to have a 'sealed packet' (as far as Workstep is concerned, a variable length string) for data that is simply passed through and has no effect on Workstep.

The general form of an invocation message would then be:

- invoked object class
- instance identification items
- message type (event/super event type) and instance
- data items issued by this object class
- group of items each containing:
  - object class
  - 'sealed packet' (variable length string).

6.4.3 Response messages

In many cases the response message will simply be 'success' or 'failure', but sometimes data is needed by the invoking object. For example, suppose that Contract - work contained a daily rate, needed to work out the cost of the assigned Task, so that Task Cost could be set against Workstep Budget. There are two possibilities:

- if Task had a Cost attribute, Contract Rate would be response data sent to Task, and Task Cost would be response data sent to Workstep

- if Task did not use Contract Rate, but simply passed it through to Workstep, the 'sealed packet' convention would be used.

6.4.4 External output

As well as responding to its invoked object, an object may need to provide output to the External Design. The simplest way to handle it is to send it with the response message; this is another example of 'passed-through' data that would require a 'sealed packet' convention.

An alternative would be to send it directly to the event manager (using the event instance identifier), as illustrated in Figure 6.7. This shows Workstep Closure, producing a report summarising what work had been done, which Tasks were completed and which were cancelled before completion, etc.

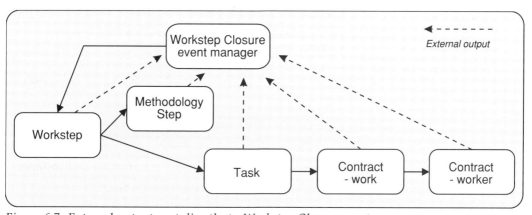

Figure 6.7: External output sent directly to Workstep Closure event manager

There is a complication for the event manager with this approach. There are an indefinite number of tasks for a workstep, and a contract for each task. The event manager needs to know when it is finished – ie when it has received the output data from all affected instances of Task, Contract - work and Contract - worker. One way of managing this is for Task to send its output as a single batch, rather than as individual output messages for each affected instance. The event manager could then match Contract responses against Tasks in the batch.

6.4.5 Combined response

If the external output is to be passed back through the network, rather than directly to the event manager, we need to specify messages that contain both types of output:

- data items needed by the invoking object class

- group of items each containing:

  – object class

  – 'sealed packet' (variable length string).

One of the object classes for which sealed packets are transferred is the event manager. If invocation and response is to be handled by a message-handling software layer (rather than by, say, invocation of stored procedures), then routing information will also be needed.

6.4.6 Nested structure

Iterations and selections in the full ECD will require a nested message structure. For example, the message received from Workstep by the Workstep Closure event manager would be something like:

> success/failure status
> Workstep Description
> (Methodology Step Description)
> indefinite number of Task packets, each one
>   either:
>   ('cut-short', Task Number, Task Description, Hours Worked, Payment Made, (Contract Id, 'last task?' status))
>   or:
>   ('completed', Task Number, Task Description, (Contract Id, 'last task?' status))

Parentheses indicate sealed packets.

6.4.7 The event manager

The event manager is fairly simple:

- get message from External Design
- invoke Workstep using invocation message
- unpack Workstep response into form expected by External Design
- return event response to External Design.

6.4.8 Two-phase commit

A two-phase commit is needed. Each correspondence arrow on the diagram implies an invocation and a response. The event manager will issue a commit instruction only when it has had a 'success' response for the network of invocations.

6.5 Implications of using direct invocation

With direct invocation the event manager has responsibility for communication between the object classes in the ECD. The appeal of this approach lies in its apparent simplicity:

- related objects do not need to know about each other; this knowledge is moved into the event manager
- there is a simple, two-phase commit rather than the distributed commit in the network approach.

For example, Figure 6.8 shows direct invocation of object classes for Assignment of Task to Contract.

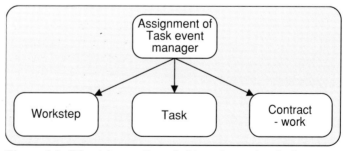

Figure 6.8: Direct invocation of object classes for Assignment of Task to Contract

6.5.1 Super-events

Re-usability is preserved. We have to define an event handler for each super-event. Event managers can then invoke entity-based object classes and other event managers, as illustrated in Figure 6.9, for Project Closure.

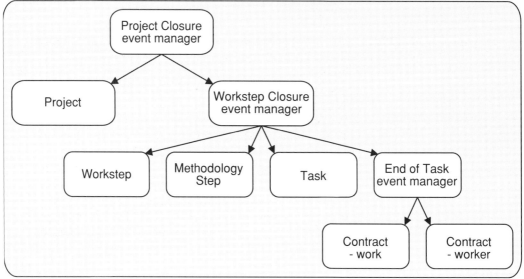

Figure 6.9: Reuse with direct invocation for Project Closure

6.5.2 Response data

Message-passing becomes more complex if response data is needed by the invoking entity.

In the example of Assignment of Task to Contract, used in Section 6.4, Contract Rate is needed to calculate Task Cost, and Task Cost is needed to establish whether Workstep Budget can afford the assignment. Figure 6.10 shows the messages needed.

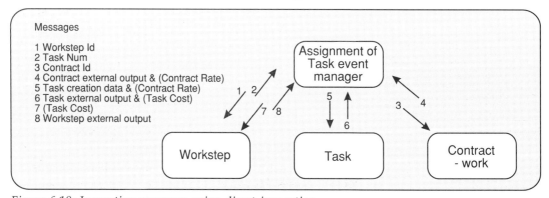

Figure 6.10: Invocation responses using direct invocation

6.5.3 'Sealed packet'

The 'sealed packet' convention indicated by parentheses in the messages in Figure 6.10, is needed for invocation response data. The event manager should not have to be changed if the data passed between two of the invoked entity-based object classes changes.

6.5.4 Fragmented methods

Workstep's method for Allocation of Task to Contract has to be split into two parts. This means that there will be two invocations of the same Workstep instance for the event. Both will update Workstep – one incrementing Task Count, the other adjusting Workstep Budget. The Workstep object class has to be able to cope with this. This issue is revisited in Internal Design, in Chapter 7.

6.5.5 Nested structure

As ECDs become more complex and correspondences occur lower down the structure, it becomes more difficult to delegate object-to-object communication to the event manager. For example, the ECD for Workstep Closure and the invocations needed are illustrated in Figure 6.11.

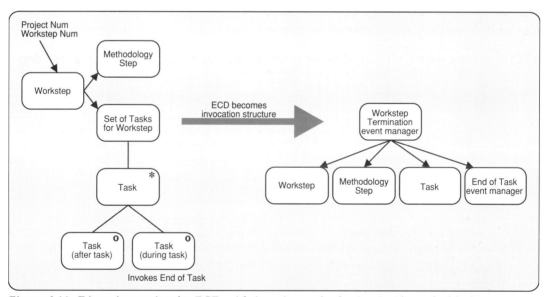

Figure 6.11: Direct invocation for ECD with iteration and selection inside method body

ECD substructures are hidden as method bodies inside object classes. However, to produce an event manager we need to know about effect-to-effect (or effect-to-super-event) correspondences.

The effect of Workstep Termination on Task is a class method (since the event manager knows only the Workstep identifier, not the identifiers of the individual Tasks affected).

The response from the Task object class will be a list of responses for affected Task instances. The event manager must then decide which instances of Task require invocation of §End of Task (a super-event that affects Contract).

A structure for the event manager is given in Figure 6.12. What we have had to do is replicate some of the structure of the method bodies in the event manager.

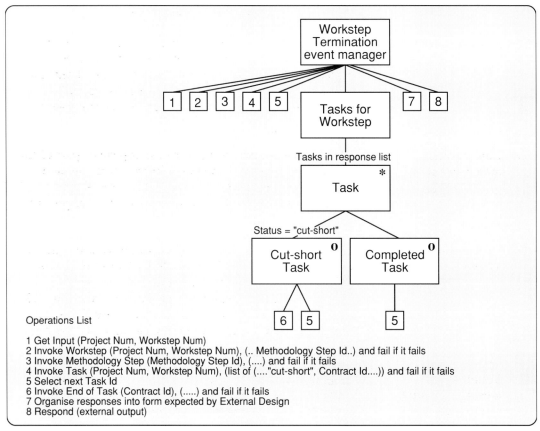

Figure 6.12: Structure of event manager for Workstep Termination

There are two points to note about this replication:

- we could optimise individual examples by re-coding them – for this one the Task object could produce two lists of task instances, 'completed' and 'cut-short', so that the event manager would need only an iteration of 'cut-short' tasks and not the selection – but there is no general optimisation. Also, this kind of optimisation distorts the specification derived from the ELHs and ECDs and we still entertain hopes of generating much of the required code from them

- complexity increases substantially without super-events; with super-events, some of the replication is contained within event managers.

6.5.6 Batch processing

Where there are nested structures of iterations, event managers for direct invocation may have to process responses that are, in effect, batches of transactions. For example, when a project is closed there is a set of Worksteps for the Project, and a set of Tasks for each Workstep.

6.5.7 Configuration management

When there are two copies (or near copies) of part of the processing structure – one distributed across method bodies, the other in an event manager – there is a requirement to keep them consistent when changes are made.

6.5.8 Update Process Model

The event manager for direct invocation is fairly closely related to the Update Process Model (UPM) of *SSADM V4 Reference Manuals*. One technique for creating an event manager (especially useful if there is a CASE tool available that will generate UPMs automatically from earlier products) is to create the UPM and edit the event manager from it.

6.6 Summary of message-passing approaches

Both of the event manager approaches are useful, in different project situations. Figure 6.13 illustrates the types of ECD suitable for each approach.

It is generally simpler to use one or other approach for a whole system but, in theory, both types of invocation could be used, provided that the implementation software can handle the commits.

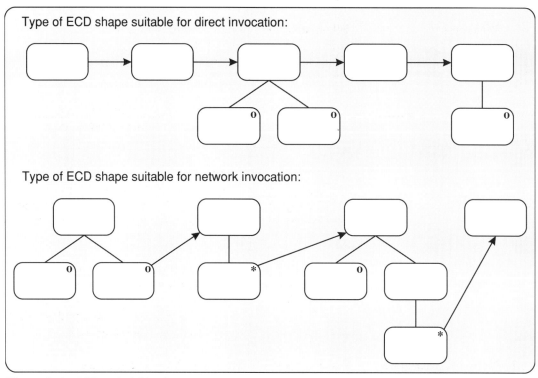

Figure 6.13: ECD suitability for message-passing approaches

6.6.1 Direct invocation

Direct invocation would be suitable for systems where:

- there is relatively simple state modelling in the system – eg record-keeping systems where most entities have only three possible states, 'not active yet', 'active' and 'history' – and most ECD correspondences are at the top level

- the database is to be stored at one location.

Strengths

Its strengths are:

- simple design if most ECD structures are 'flat'

- generally, messages are small and simply-structured

- less dependence on the 'sealed packet' convention (since invocation response data is less common than external output)

- basic two-phase commit is needed.

Weaknesses		When state modelling is richer, and more correspondences occur between components within lower-level iterations and selections, direct invocation becomes more cumbersome:

- structure of method bodies is partly replicated in event managers (and changes may have to be replicated)

- the conversion procedure from ELHs and ECDs to object class managers, method bodies and event managers is less direct – this lowers traceability and lowers the possibilities for automated support for the transformation

- there are large numbers of small messages, compared with network invocation. This will be of concern for distributed systems – communications may become slower and more expensive (unless we partition event managers between locations).

6.6.2 Network invocation

For entity-events with more knowledge (rules, constraints, state management) built into them than basic record keeping, the network approach is generally a better solution than the direct approach.

Strengths

Its strengths are:

- it works cleanly in all situations, however complex, without structure replication

- simple structures for event managers

- there is a direct transformation of products into coding specifications, and hence much stronger possibilities for automatic generation than with direct invocation.

Weaknesses

The main weaknesses of network invocation is that for some projects it will be overkill to create elaborate nested message structures. A large proportion of projects are basic record-keeping systems, with very limited state modelling. The weaknesses are:

- possibly complex structures of 'sealed packets' to be passed between object classes

- distributed commit is needed.

6.6.3 Event manager is part of conceptual model

Note that whichever approach is used, the event manager is part of the Conceptual Model (even if, when implemented, it is located on a client in a client/server architecture).

7 External and Internal Design

7.1	Introduction	This volume concentrates on the specification of the Conceptual Model. This chapter briefly discusses some External and Internal Design issues.
7.2	External Design	External Design objects, such as buttons and windows, are not a prerequisite for using the notion of objects in the Conceptual Model or Internal Design. Figure 7.1 illustrates how the External Design is linked to the Conceptual Model.

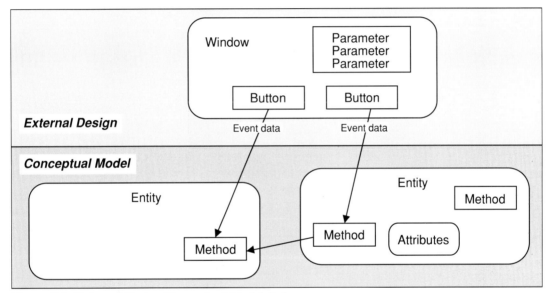

Figure 7.1: External Design/Conceptual Model links

The External Design and Conceptual Model are clearly separated in SSADM. OO authorities agree, arguing that separate object models must be produced for the External and Conceptual Schemata.

As indicated in Chapter 3, the method-specification approach to OO design may be most appropriate to the External Design. It seems likely that OO ideas to do with composite objects and inheritance will be more useful.

Composite objects — User interface objects such as windows and reference lists certainly look like composite objects. Using some GUI design tools, the ease of creating objects which inherit the properties of previously designed objects, enables designers to put together new user interface designs swiftly. This is especially attractive where the early presentation of a prototype design to users is expected or needed.

Inheritance — User interface designers use inheritance in an informal way all the time. Modern user interface management systems enable them to copy an object from one window or dialogue to another (carrying forward the properties of the object as they do this) then they revise the new object. This is an intuitive, temporary, even playful, use of inheritance, however, it can be formalised by developing standards and style guides to impose the reuse of design components.

Event/enquiry managers — There is clear correspondence between the 3-schema Specification Architecture and client/server implementation. The simple mapping (eg for first-cut design) is:

- External Design to workstation clients

- event and enquiry processes to server processes.

If event/enquiry manager components are used to invoke object classes and co-ordinate output, there may be architectural and performance advantages in placing the implemented event/enquiry managers on workstation clients, as illustrated in Figure 7.2.

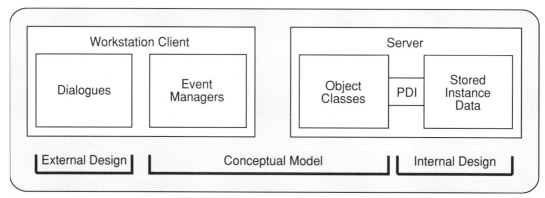

Figure 7.2: Possible mapping of specification components to client/server design

Note that placing event managers on workstation clients does make them part of External Design. They are controlled and managed as part of the Conceptual Model. They do not change if I/O technology changes, or if events and enquiries are grouped differently into functions. They are re-usable by different External Designs.

7.3 Schema obligations

A major theme of SSADM is the need to simplify the jobs of system specification and maintenance by separating concerns. The SSADM rationale promotes the 3-schema Specification Architecture as a device for doing this. Of course, the three schemata cannot entirely be separated; they must be designed in harmony and there must be communication between them.

Cross-schema obligations

Within the scope of a system being designed, input data should be validated on entry to the system, and validation tests should not be repeated thereafter. In terms of the 3-schema Specification Architecture, each schema has an obligation to provide valid data to the others.

'Valid' here means at least correct in terms of format and syntax. The External Design has an obligation to invoke the Conceptual Model with syntactically correct events. The Conceptual Model has an obligation to invoke the Internal Design with syntactically correct operations such as create, read, write and delete entity. In the same way, the responses to these invocations should be syntactically correct.

The designer of each schema should have faith in the designer of the other schemata. The alternative is anti-reuse; it means duplication in both the system specification and the implemented code. This can give problems in specification, maintenance, configuration management and system performance.

Within-schema obligations

Likewise, within each schema, each process or object must be responsible for doing its job correctly and must output only valid data to the next.

7.4 Database implementation mechanisms

We may pass to the database designer an LDM which contains both class hierarchies and aggregations. In database design, several possible implementation mechanisms have been established. In Chapter 5 we used the LDM shown in Figure 7.3.

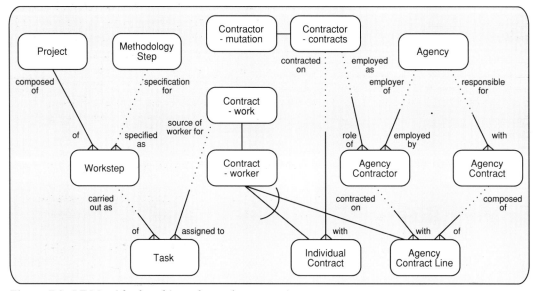

Figure 7.3: LDM with class hierarchy and aggregation

The LDM in Figure 7.3 contains a class hierarchy for contract and an aggregation of parallel aspects for contractor. As was discussed in Chapter 5, for a class hierarchy to be useful in implemented systems, the super-type must provide common behaviour (methods) that can be inherited as well as common attributes. We model the class hierarchy as an aggregation of one generalised behaviour component (Contract - work) and one specialised behaviour component (Contract - worker).

7.4.1 Simple aggregations

The Contractor aggregation has two aspects: Contractor - contracts, which manages the one-contract-at-a-time behaviour, and Contractor - mutations, which maintains the contractor personal data. There are two options for implementation: one table or separate tables.

One table	The default option is to implement all aspects of the aggregation in one table, with each aspect defined as a view. The implemented object classes can remain distinct. Since the views are single-table views, update processes will work without change. The only problem is creation and deletion of rows. The birth and death processes will (at the conceptual level) create and delete operations in the event processes, it is preferable to deal with them in the PDI. This keeps the processes in the Conceptual Model robust if the implementation is changed to separate tables, as discussed next.
Separate tables	The alternative is to implement aspects in separate tables, co-ordinated by a common key. This option is worth considering if:

- the aspects are to be stored in different places

- one of the aspects is in a shared server, used by other systems

- one of the aspects is an 'object-wrapped' legacy system

- the system is partitioned on aspects, and subsystems are to be delivered at different times with minimum disturbance to already-installed databases. This approach is described in ISE Library volume *Application Partitioning and Integration with SSADM*

- there is a worthwhile performance improvement, for example if one aspect is much smaller than the other, and used much more frequently.

There will be higher numbers of physical updates when both aspects are affected by the same event.

7.4.2 Class hierarchies

In a relational implementation, if database code is to be inherited from a super-type, all instances of the super-type must be in the same table.

The basic options for implementing the class hierarchy are:

- the entire class hierarchy in one table

- each sub-type and super-type in a separate table, as illustrated in Figure 7.4.

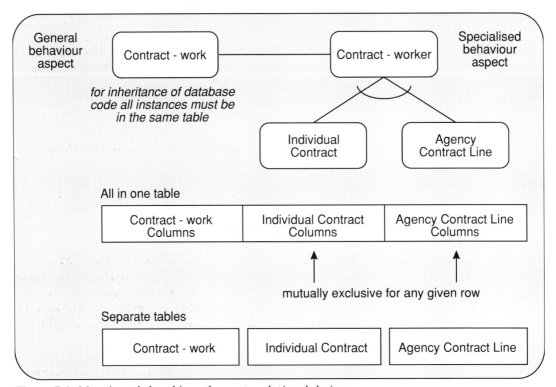

Figure 7.4: *Mapping of class hierarchy on to relational design*

One table — As for the simple aggregation in Section 7.4.1, the entire class hierarchy - general and specialised aspects, for any number of levels - can be mapped on to one table. The aspects can be implemented as if they were distinct object classes, but the instance data is defined as views of the table. The potential problem is wasted space (or processing time for compressed formats) for large tables with mutual exclusion defined for large numbers of columns.

Separate tables — There may be space or performance savings in storing the super-type and each sub-type in a separate table. The database code for the super-type is inherited.

The penalty (in addition to higher numbers of physical updates) is that the object class for specialised behaviour has to deal with two tables. If it is retrieving by object identifier rather than user-friendly key, it may have to query both tables to find the required instance.

Sub-types in one table?

Generally, there is no point in implementing both sub-types in one table and the super-type in another. Both sets of overheads would be incurred – additional physical updates and mutually exclusive columns.

Rare exceptions might be:

- if sub-types and super-type attributes were to be stored at different locations

- if the super-type were to be implemented as a shared server, whose behaviour was to be inherited by sub-types in other applications.

Rolling down the super-type

There is a view held by some OO practitioners that super-types are abstract, and objects are instantiated only at the sub-type level. Unless the entire class hierarchy is constrained to be implemented in one table, there would in some projects be a need to implement separate tables for each sub-type.

The implications of this, as illustrated in Figure 7.5, are:

- the event/enquiry processes for the super-type could be inherited, provided that the PDI hid the separate tables from the Conceptual Model

- the database code would not be re-usable.

7.4.3 Selective locking

Relational database management systems generally provide locking, commit and rollback at the transaction level. However, the approach described in this volume may result in transactions that are too fragmented to be recognised as a single transaction by the DBMS.

This would mean that each object class invoked for an event would have to declare its own transaction to the DBMS for each event instance, and then issue the database commit (or rollback) when it received the appropriate message from the event manager.

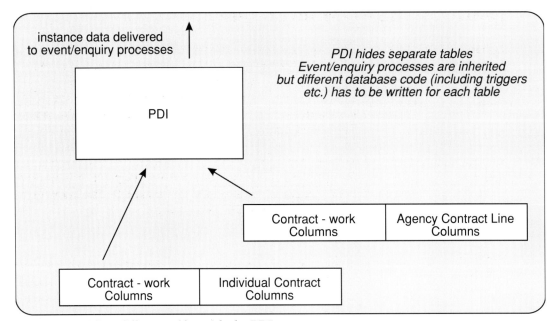

Figure 7.5: Hiding different tables with the PDI

Some object instances might be affected more than once by an event instance. For example, when a Project is cancelled, a Contract might have more than one active Task on it. The Contract object class would be invoked (directly or indirectly) by each 'cut-short' active Task. On the first invocation, the object class would read the Contract instance and lock it. When it received subsequent invocations for the Contract instance, it must be able to detect:

- additional invocations for the same event instance, and permit the update

- invocations for a different event instance, and refuse them.

The invocation message must include an event instance identifier, as well as the message type.

8 Impact on SSADM techniques

8.1 Introduction

The extensions to SSADM recommended in this volume affect the use of existing techniques. No changes to the default SSADM structure have been necessary to accommodate them. However, the extensions to the techniques do impact the detailed activities within some SSADM steps.

8.2 Impact on SSADM steps

This section summarises the impact of the extensions described in this volume on SSADM steps. It also provides cross-references to sections in earlier chapters describing the modified techniques.

Step 210: Define Business System Options

Unchanged, but:

- specify the 'identity-based' view of the LDM (Section 4.5).

Step 320: Develop Required LDM

Largely unchanged, but:

- specify the 'type-based' view of the LDM looking for common components (Section 4.5)

- create a first-cut event catalogue (Section 5.4).

These prepare the ground for later analysis steps. Common components (super and sub-type entities) are identified in the LDM, even though they may be discarded during later steps.

Step 330: Define System Functions

There are two minor modifications:

- define common domains for I/O data items (Section 3.8.1)

- define common elements in data flow structures (Section 3.8.2).

Step 340: Enhance Required LDM

Unchanged, apart from:

- map entity class hierarchies on to TNF relations (Section 4.6).

This means studying all the super and sub-type entities, looking to map them on to the results of RDA.

Step 360: Develop Processing Specification	Enhancements to processing specification include:
	
	• separation of parallel aspects
	   – specify the 'object' view of the LDM (Section 4.5)
	   – define parallel lives in separate ELHs (Section 5.6)
	
	• specification of processing detail on ELHs
	   – use revised operation syntax (Annex C)
	   – specify extra preconditions as 'Fail' operations (Section 5.2.2)
	   – optimise state-indicator values (Section 5.3)
	   – recognise common effects in ELHs as super-events (Section 5.5.2)
	   – construct event class hierarchy (Section 5.4.2)
	
	• enhanced specification of ECDs
	   – construct ECDs with one-many correspondence arrows (Section 5.7)
	   – indicate invocation of common ECD for super-event (Section 5.7).

Step 520: Define Update Processes	Unchanged, but: • use revised operation syntax (Annex C) • specify each event using the message-passing and procedural paradigm (5.7).

Step 620: Create Physical Data Design	There are two modifications to this technique:
	
	• implement each class hierarchy (Section 7.4) using either:
	   – one table
	   – separate tables
	   – sub-types in one table.
	
	• implement each aggregation (Section 7.4) using one of:
	   – one table
	   – separate tables.

Step 660: Consolidate PDI	Unchanged, but: • shield the entity-event model from database implementation variations (Section 7.4).

A The System Development Template and the 3-schema Specification Architecture

A.1 Introduction

This annex briefly describes the System Development Template and 3-schema Specification Architecture. Any reader requiring further information on either of these concepts should refer to the ISE Library volume *Customising SSADM*.

A.2 The System Development Template

The *System Development Template* (SDT) provides a common structure for the overall system development process. The 3-schema Specification Architecture, described in Section A.3, concentrates on those products which will ultimately lead, sometimes via other products, to elements of software. The SDT takes a broader view. It divides the development process into activity areas and maps products on to those areas. The SDT is shown in Figure A.1.

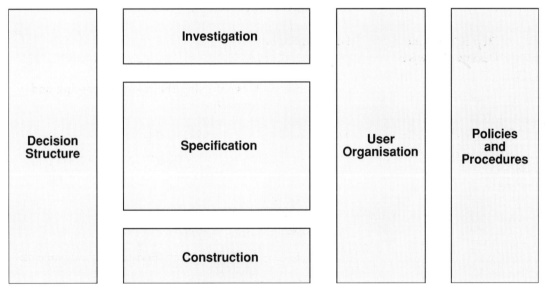

Figure A.1: The System Development Template

A.3 The 3-schema Specification Architecture

The *3-schema Specification Architecture* is a model which divides a system design into three areas or 'views':

- an External Design which comprises the user interface

- a Conceptual Model which comprises the essential business rules and knowledge

- an Internal Design which defines the physical database design and the process/data interface (PDI).

The 3-schema Specification Architecture helps maintain a level of independence between the model of the essential business requirements, the user environment and the physical system implementation. It also helps to partition the system design between its logical and physical components. A change to the technical environment should only require changes to the Internal Design and the physically-dependent elements of the External Design, such as physical dialogue design or changes arising from the appearance of a display image.

Hence the 3-schema Specification Architecture facilitates changes arising from physical and environmental factors and increases the robustness of the system design. System enhancements are more cost-effective since a change to any one element will require a minimum of changes to other elements.

SSADM can be mapped on to the 3-schema Specification Architecture and both can be mapped on to the SDT. Figure A.2 shows the 3-schema Specification Architecture mapped on to the SDT.

Annex A
The System Development Template and the 3-schema Specification Architecture

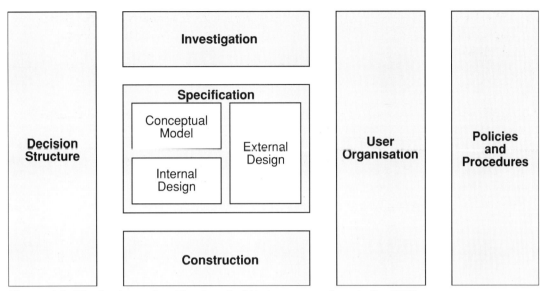

Figure A.2: The 3-schema Specification Architecture mapped on to the SDT

A.4 Main area of impact of OO approach in this volume

The main impact of the OO design approach described in this volume is on the 3-schema Specification Architecture in the System Development Template. See Figure A.3.

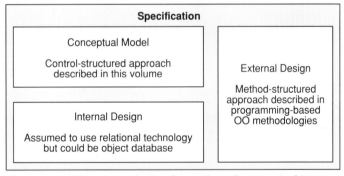

Figure A.3: Impact on the 3-schema Specification Architecture

137

B Specification for Contractor case study

B.1 LDM for Contractor case study

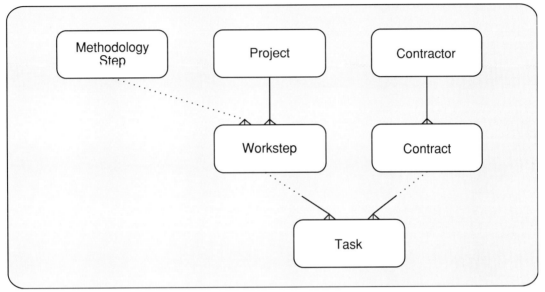

Figure B.1: LDM for Contractor case study

Notes on 'at least one' relationships	The 'at least one' rule appears on two relationships in the data model in Figure B.1. Many database management systems cannot implement this rule. There are at least two varieties of the rule which can be distinguished by specification in the entity-event model.
	First, we can specify the rule that one detail is created on the birth of the master. In the case study, the Project Opening event must carry one (and only one) Workstep with it.
	Second, we can specify the rule that the master is created on the birth of the first detail. In the case study, the rule that an Employee is created along with the first Employment Contract.
	In both cases, the implementation of the rule as a 'fail' statement can be automatically generated in the ECD or UPM, from ELHs.

Entity	Attributes	How attribute values change
Contractor	<u>NINo</u> Contractor Name Contractor Address Contract Num (next)	replaceable by Change of Name replaceable by Change of Address incremented by Contract Agreement
Contract	<u>NINo</u> <u>Contract Num</u> Active Tasks	incremented by Assignment of Task decremented by Completion of Task, Workstep Closure and Project Closure
Methodology Step	<u>Methodology Step Num</u> Methodology Description*	stored on Workstep Definition stored on Workstep Definition
Project	<u>Project Num</u> Project Description* Workstep Num (next)	stored on Project Opening incremented by Workstep Definition
Workstep	<u>Project Num</u> <u>Workstep Num</u> Methodology Step Num Task Num (next)	stored on Workstep Definition incremented by Assignment of Task
Task	<u>Project Num</u> <u>Workstep Num</u> <u>Task Num</u> Task Description* NINo Contract Num Hours Spent Payment Made	stored on Assignment of Task foreign key foreign key incremented by Record of Hours Spent incremented by Record of Payment Made
* This attribute defines the entity itself and could be considered as a candidate key for the entity. The user will not be provided with the means to alter this attribute.		

Figure B.2: Entity catalogue for Contractor case study

B.2 Event Catalogue

Events have been categorised in the following tables as birth, death, attribute update and deletion events.

The first group of events listed in Figure B.3 are birth events.

Event	Event Data	Event Effects	Event Constraints
Project Opening	Project Num *[The next available Project Num is supplied automatically].* Project Description Methodology Step Num	Unlike most birth events, which create only one instance of one entity type, this event will create not only a Project with a Project Description but one Workstep with a Methodology Step Num. Workstep Numbers must be allocated to each Workstep within the Project starting from 1, and the Workstep Num (next) must be recorded in the Project entity.	Project must not exist. Methodology Step Num must be valid.
Addition of Workstep to Project	Project Num Methodology Step Num Workstep Num *[The relevant Workstep Num is not input, but taken from Workstep Num (next) in the Project.]*	This event will add a Workstep to an existing Project and set Task Num (next) = 1.	Project must exist. Methodology Step Num must be valid.
Assignment of Task to Contractor	Project Num Workstep Num NINo Task Description Task Num *[The relevant Task Num is not input, but taken from Task Num (next) in the Workstep.]*	This event will add a Task to an existing Workstep, assigning it to the current Contract of a Contractor.	Workstep must exist and not be finished. Contractor must have a current and active Contract.

Event	Event Data	Event Effects	Event Constraints
Contract Agreement	NINo [Contractor Name] [Contractor Address] Contract Num [Since the new Contract must be the next, the relevant Contract Num is not input, but taken from Contract Num (next) in the Contractor.]	This event will add a new Contract to a Contractor and increment Contract Num (next). If this is the first Contract for the Contractor, then it will create a new Contractor as well. This means the input data must contain optional data, marked here with square brackets. (Since the name and address are mandatory input on the first Contract Agreement, the user would recognise the first Contract Agreement as a different kind of event. However, we want an example in the case study where the birth of the first detail automatically creates a master entity.)	Contractor must exist. Contractor must not already have a current active Contract.
Methodology Step Definition	<u>Methodology Step Num</u> Methodology Description	This event will create a Methodology Step.	Methodology Step must not exist.

Figure B.3: Event Catalogue for birth events

The events listed in Figure B.4 are death events. Assume that none of these death events will delete any entity. Deletion is considered in Figure B.6.

Event	Event Data	Event Effects	Event Constraints
Project Closure	Project Num	This event will freeze the Project entity, preventing any further update events from affecting it. It will cascade down to have a similar affect on all of its Worksteps and Tasks.	Project must exist and not already be closed.
Workstep Closure	Project Num Workstep Num	This event will freeze the Workstep entity, preventing any further update events from affecting it. It will cascade down to have the same affect on all Tasks.	Workstep must exist and not already be terminated.
Completion of Task	Project Num Workstep Num Task Num	This event will partly freeze a Task in that no more time can be recorded against it, However, further payments to the Contractor can be recorded after this event.	Task must exist and not already be terminated.
Contract Termination	NINo Contract Num *[Since only the latest Contract can be terminated, the relevant Contract Num is not input, but deduced by subtracting one from Contract Num (next) in the Contractor.]*	This event will partly freeze a Contract entity, in that no more Tasks can be assigned to it, However, it will have no affect on active Tasks, and further Task Completions can be recorded after this event.	Contractor must exist. Contract must exist and not already be terminated.

Figure B.4: Event Catalogue for death events

The events listed in Figure B.5 are attribute update events.

Event	Event Data	Event Effects	Event Constraints
Contractor Address Change	NINo Contractor Address	This event will replace the stored Contractor Address with the one input.	Contractor must exist.
Contractor Name Change	NINo Contractor Name	This event will replace the stored Contractor Name with the one input.	Contractor must exist.
Record of Hours Spent	NINo Task Num Hours Spent Contract Num *[Since only the latest Contract can have time recorded against it, the relevant Contract Num might not be input, but be deduced by subtracting one from Contract Num (next) in the Contractor.]*	This event will add the time input to the time already recorded.	Task must exist and not already be completed.
Record of Payment Made	NINo Task Num Payment Made Contract Num *[Since the Contractor can only be paid for the latest Contract, the relevant Contract Num might not be input, but be deduced by subtracting one from Contract Num (next) in the Contractor.]*	This event will add the payment made to the payment already recorded.	Task must exist and not already be terminated by project or Workstep closure.

Figure B.5: Event Catalogue for attribute update events

The final group of events listed in Figure B.6 are deletion events, for which a deletion strategy must be designed. Task, Workstep and Contract entities cannot be deleted by themselves (this would make any report starting from Project or Contractor incomplete). Deletion can only be done from the top downwards, under very controlled conditions as described next.

Event	Event Data	Event Effects	Event Constraints
Contractor Deletion	NINo	This event will delete the Contractor and all his/her Contracts. It will cut each of the Tasks from the deleted Contract.	The latest Contract must be finished (that is, all Worksteps for the latest Contract must have been completed or otherwise terminated).
Project Deletion	Project Num	This event will delete the Project, all its Worksteps and all its Tasks.	Project must be closed. Either no Task ever assigned to the Project, or all Tasks on all of the Worksteps must have been cut from their Contractor by Contract Deletion.

Figure B.6: Event Catalogue for deletion events

Note that the effect of Contractor Deletion is to make the relationship from Task to Contractor optional. Note also a potential partition of the LDM as shown in Figure B.7. See the ISE Library volume Application Partitioning and Integration with SSADM for guidance on dividing an LDM between the parallel aspects of an entity.

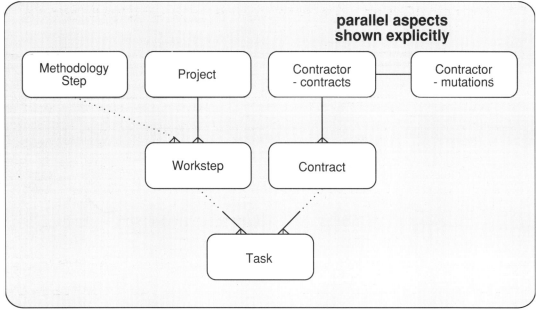

Figure B.7: Parallel aspects of Contractor shown explicitly on the LDM

Annex B
Specification for Contractor case study

B.3 Remaining ELHs for the case study

This section provides the remaining five ELHs for the Contractor's case study.

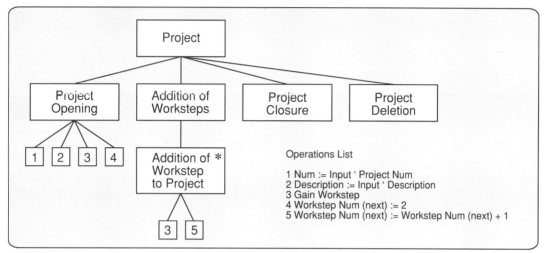

Figure B.8: Project ELH

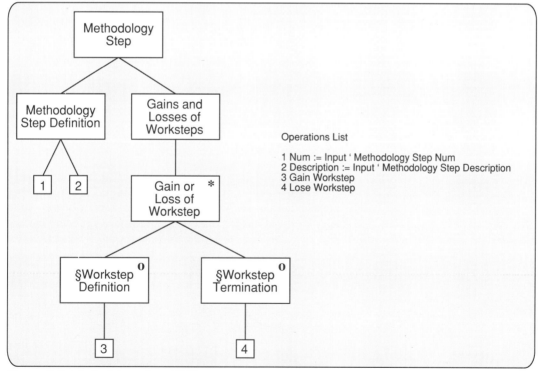

Figure B.9: Methodology Step ELH

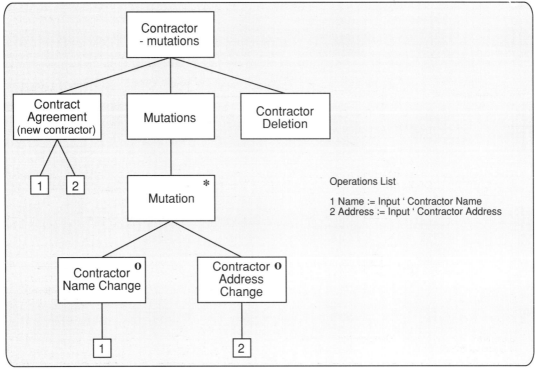

Figure B.10: Contractor - mutations ELH

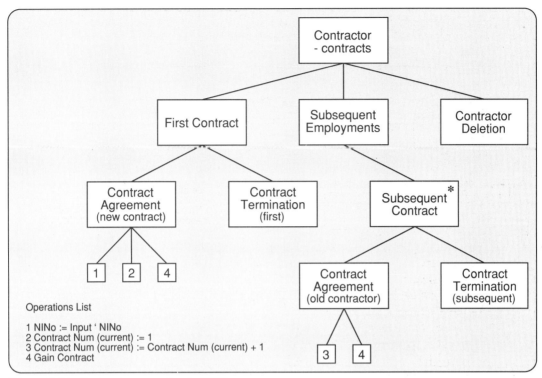

Figure B.11: Contractor - contracts ELH

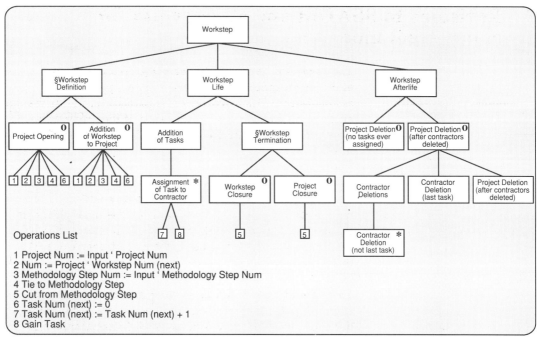

Figure B.12: Workstep ELH

C Revisions to SSADM operation syntax for entity-event modelling

The operation syntax in the SSADM V4 manuals for entity-event modelling is revised below, partly to accommodate OO extensions and partly to assist in the automatic generation by CASE tools of these operations from information specified in the LDM and entity-event model.

Abbreviations

In all entity and attribute operations, the <entity name> may have to be more fully expressed as <entity name>[<aspect name>][<role name>]. The aspect name is the name of the parallel ELH in which an event appears. The role name distinguishes between different kinds of the same entity type affected by one event.

Throughout 'value' can be an item name, a literal or an expression. Item names and expressions are not further defined here, but an item name would normally be an attribute name.

C.1 Entity operations

New or modified entity operations are:

- Write <entity name>

- Delete <entity name>
  - A delete is a kind of write (last)

- Read [next]<entity name>[<relationship name>][where selection condition][if absent set<entity name state indicator>=null [and create <entity name>]]
  - [next] – used for reading through sets
  - [where selection condition] – used for direct access on a key and for skipping irrelevant details when reading around a set
  - [if absent set <entity name state indicator> = null] – helpful because it enables the state-indicator value to be tested in later iteration and selection conditions
  - [and create <entity name>] – used only on birth events creating an entity

C.2 **Relationship operations**

New or modified relationship operations are:

- Tie to <master entity name>[<relationship name>]

- Cut from <master entity name>[<relationship name>]

Where cut and tie operations are paired together, they are combined thus:

- Swap <master entity name>[<relationship name>]

- Gain[set of]<detail entity name>[<relationship name>]

- Lose[set of]<detail entity name>[<relationship name>]

Allocating gain and lose operations to event-effects in an ELH ensures that the necessary 'validation effects' of events are documented, and reflects the updating of indexes which is necessary in most implementation environments. Gain and lose operations may not have to be implemented in a non-distributed relational database environment.

C.3 **Attribute operations**

New or modified attribute operations are:

- <item name>[<role name>]:=value

    - this single assignment operation replaces all 'store' and 'replace' operations in the SSADM v4 manuals; <role name> is used to distinguish between different versions of the same attribute within one entity.

- Set <entity name>state variable = value

    - a variant of the assignment operation for state variables only.

- Fail unless <expression>

– this fail operation would normally test the value of a state indicator, but it may be added for any other kind of precondition.

C.4 Other data model processing operations

Other new or modified data model processing operations are:

- Invoke <subroutine name> and fail if it fails

  – the Invoke operation can be used to call any common routine such as a calculation, or super-event.

- Checkpoint<checkpoint name>

- Restore<checkpoint name>

Checkpoint and Restore operations may have to be allocated where there is a recognition problem in a UPM.

C.5 Message file operations

New or modified message file operations are:

- Open <file name>

- Close <file name>

- Put <file name><record name>

- Get<file name>

A Get operation can be allocated to collect the event at the start of a UPM, and to read through any iterated data groups within the event data.

D SSADM V4 concepts and corresponding OO concepts

The table in Figure D.1 summarises the closest correspondences between conceptual modelling concepts in SSADM V4 and OO concepts.

SSADM V4 Concepts	Corresponding OO Concepts
Entity Life History Super- and sub-types Attribute Relationship	Object Class hierarchy (inheritance tree) Attribute (instance variable) Association
Entity aspect[1] Entity Life History Valid prior state Effect Operation allocation Event Super-event[3] Effect Correspondence Diagram Selection of optional effects triggered by event on entity	Object[2] Object as finite state machine Pre-condition Method[4] Method body step Transaction Re-usable set of transactions Aggregate object[2] Method

Notes:
[1] idea to be extended for reuse benefits or better correspondence with oo ideas
[2] strong correspondence, but perhaps not intuitively obvious
[3] new idea, added for reuse benefits and better correspondence with oo ideas
[4] weak correspondence, though usually true in simple cases

Figure D.1: Closest correspondences between OO and SSADM V4 concepts

Bibliography

SSADM V4 Reference Manuals

The SSADM Version 4 Reference Manuals were produced by CCTA. They are published by and available from NCC Blackwell Ltd, 108 Cowley Road, Oxford, OX4 1JF.

ISBN 1 85554 004 5.

Information Systems Engineering Library

The Information Systems Engineering Library volumes are available from the HMSO Publications Centre. Telephone orders may be made on: 0171 873 9090. Full details of how to obtain HMSO publications are provided on the back cover of this volume.

ISE Library volumes referred to in this volume are:

- Customising SSADM
  ISBN 0 11 330664 4

- SSADM and GUI Design - A Project Manager's Guide
  ISBN 0 95 221530 6

- Application Partitioning and Integration with SSADM
  ISBN 0 11 330622 9

- Distributed Systems: Application Development
  ISBN 0 11 330623 7

- An Introduction to Reuse
  ISBN 0 11 330625 3

- Managing Reuse
  ISBN 0 11 330616 4

Object-Orientation

The following books on Object-Orientation were referenced in this volume:

Object-Oriented Modelling and Design
Rumbaugh, Blaha, Premerlani, Eddy and Lorensen, Prentice Hall (1991), ISBN 0 13 630054 5

Object-Oriented SSADM, Robinson & Berrisford, Prentice Hall (1994), ISBN 0 13 309444 8

Object-Oriented Software Construction
Meyer, Prentice Hall (1988), ISBN 0 13 629049 3

Glossary

3-schema Specification Architecture	A framework for IT system specification and implementation in which the IS services needed to support business activities are defined in a Conceptual Model. The mappings from the Conceptual Model to specific user organisations and implementation technologies are defined in an External Design and an Internal Design. In SSADM V4 the concepts of the 3-schema Specification Architecture are embodied in the Universal Function Model.
aggregation	An object-oriented term describing classes which are composed of (or consist of) other classes.
aspect	An application's view of a real-world entity type, which models part of the behaviour of the entity type. What is usually called 'an entity' on an LDM is really an aspect, since it represents only a partial view of the world entity type. In most cases a real-world entity type has only one LDM aspect; there is no confusion in referring to 'LDM entities'. Aspects have to be distinguished only when there are multiple views of the same entity type that have to be co-ordinated in some way.
attribute	A descriptive property of an object class or entity that can take different values. The values of the attribute vary both with the instance of the object class or entity under consideration and with the time at which the instance is considered.
CASE	Abbreviation for Computer Aided Software Engineering.
class	An object-oriented term for a set of object instances of the same type. A class defines the methods and the attributes of the object instances.
class hierarchy	An object-oriented term for a hierarchy of classes in which sub-types inherit the methods and attributes of their super-types. Class hierarchies are sometimes referred to as inheritance trees.
class instance	See object.

composite method	A set of methods operating on one or more object types, all triggered by the same event.
composite object	A set of parallel or nested objects sharing or inheriting the same identity.
Conceptual Model	The Conceptual Schema in the 3-schema Specification Architecture, called a *model* to emphasise that it is a partial model of the business, developed by a process of analysis and discovery. It specifies the required IT services as: information support needed by business activities; data needed to provide it; processes needed to keep the data up-to-date. It also defines which business activities are to be automated. It described the services independently of the user organisation structure and the technology for implementation. In SSADM the Conceptual Model consists of the LDM, ECDs/UPMs and EAPs/EPMs, and the implemented update and enquiry processes.
Conceptual Schema	See Conceptual Model
control-structured paradigm	Is characterised by the procedure body for an object being a single block of code structured using sequences, selections (IF THEN... ELSE... constructs) and iterations (DO WHILE... constructs). This code communicates with other code only by reading and writing messages. Its major advantage is that the context in which any part of the code is executed is extremely visible.
copy reuse	Means replicating an old component in a new place (that is a new environment, function or system). 'Cloning' is another word for copying a component for use in another system and plugging it in there.
Data Flow Diagram	Shows how services are organised and processing is undertaken. It should be a simple diagram that is readily understood, so that it can act as an effective means of communication between analysts and users.
Data Flow Model	A set of Data Flow Diagrams and their associated documentation. The diagrams form a hierarchy with the Data Flow Diagram Level 1, showing the scope of the system and the lower level diagrams expanding the detail as appropriate. Additional documentation provides

a description of the processes, input/output data flows and external entities.

DBMS   Acronym for Database Management System.

DFD   See Data Flow Diagram.

DFM   See Data Flow Model.

disciplined quit   A technique for modelling recognition problems in ELHs. It is more formal than the alternative, the undisciplined quit, and provides better support for automating the derivation of ECDs and UPMs from ELHs. The major differences from the undisciplined quit are: the scope of the recognition problem must be defined by *posit* and *admit* labels; *Quit/Resume* means 'the event labelled with R(*esume*) may occur *instead of* the event labelled with Q(*uit*)'.

discriminator   A type indicator in a super-type entity which enables processes navigating the database to 'know' which of two sub-type records exists without having to search for them.

EAP   See Enquiry Access Path.

ECD   See Effect Correspondence Diagram.

Effect Correspondence Diagram   Shows all the effects an event has on data within the system and how those effects impact upon each other.

Effect Correspondence Diagrams provide the access path details for update functions which are used in Logical Design activities.

ELH   See Entity Life History.

Enquiry Access Path   The route through the Logical Data Model from an entry point to the entity, or entities, required for a particular enquiry function.

Enquiry Process Model   Consists of a structure diagram for an enquiry processing requirement and the associated Operations List. The structure is based on the Enquiry Access Path.

Entity Life History	Charts all of the events that may cause a particular entity to be changed in any way. It shows the valid structure of events (initially identified through the use of data flow modelling and function definition techniques) affecting an entity on the Logical Data Structure.
EPM	See Enquiry Process Model.
encapsulation	An object-oriented term describing the property of objects whereby the methods of any object cannot access the attributes of any other object.
event	An event is identified as whatever triggers a process (on a Data Flow Diagram) to update the values or status of the system. An event may cause more than one entity to be changed.

In the logical system, an event initiates an update process. |
External Design	The External Schema in the 3-schema Specification Architecture, called a design to emphasise that it is developed by a process of design and engineering – there is no 'right' answer. It is defined as a two-level mapping. Update and enquiry processes in the Conceptual Model are mapped on to functions that support user roles in a particular user organisation. Functions are grouped into dialogues and/or batch input/output programs that are then mapped to an input/output technology.
External Schema	See External Design.
function	A set of system processing which the users wish to schedule together, to support their business activity.
graphical user interface	A user interface which makes use of menus or graphical objects, such as icons, for selection of options and usually has a windowing capability, enabling multiple window displays on the screen at the same time.
GUI	See Graphical User Interface.
home entity	In a situation where objects need to exchange information before they can process an event, a home

entity is an entity which collates replies from objects involved in the commit unit, decides whether the event succeeds or fails, and initiates the commit phase by sending out the results of its computation.

incremental delivery An approach to development in which: releases of the implemented system are delivered within relatively short intervals; each release extends the scope of the system and is integrated with the functionality of previous releases.

information hiding A principle of information system design, whereby information not directly of relevance to the subject under consideration is suppressed or hidden.

inheritance An object-oriented term describing the property of classes which allows sub-types in a hierarchy of class types to automatically take on the methods and attributes of all their super-types, before adding methods and attributes of their own.

inheritance mechanism A specific implementation mechanism by which an OO environment can avoid storing instances of super-types (super-classes). The OO environment implements an object instance as the lowest sub-type it can be matched to, extended with properties from its super-types.

inheritance tree Synonym for class hierarchy.

instance variable See attribute.

Internal Design The Internal Schema in the 3-schema Specification Architecture, called a design to emphasise that it is developed by a process of design and engineering – there is no 'right' answer. The LDM in the Conceptual Model is mapped on to a data storage and access technology to produce a database design. Stored data is presented to implemented update and enquiry processes by a PDI, as if it were stored in the LDM.

Internal Schema See Internal Design.

I/O technology Input/output technology which is at the interface between the human and the computer system.

LDM See Logical Data Model.

Logical Data Model	Provides an accurate model of the information requirements of all or part of an organisation. This serves as a basis for file and database design, but is independent of any specific implementation technique or product.
	The Logical Data Model consists of a Logical Data Structure, Entity Descriptions and Relationship Descriptions. Associated descriptions of attribute/data items and grouped domains are maintained in the Data Catalogue.
message-passing	An object-oriented term describing the way in which objects communicate with each other.
message-passing paradigm	See message-passing.
method (object-oriented)	An object-oriented term implying the part of an object permitted to read and update specific object attributes.
method body	The specification or implementation of the main part of a method.
method step	A stage in the processing of a method.
method-structured paradigm	A paradigm for software development characterised by the implemented code's being separated into a number of loosely coupled 'methods' which are invoked by message-passing, and by the ability to classify objects and organise the object classes into a hierarchy with methods definable at any level in the hierarchy. Programs built this way make reuse of functionality easier and seem maintainable.
monogamous relationship	A one-to-many relationship between two entity types, which allows only one detail entity to be current at once (although more than one detail entity may exist over time).
polygamous relationship	A one-to-more relationship between two entity types, which allows more than one detail entity to be current at once.
procedural paradigm	See control-structured paradigm.

Glossary

object	An instance of an object class comprising both the methods of the object class and attributes with particular values.
object attribute	See attribute.
object class	See class.
operational master	An SSADM V3 term for additional entry points to logical data structures. They were used where access was required to an entity other than via its key. They were usually implemented as indexes.
operations list	A list of all the operations on either an Enquiry Process Model or an Update Process Model.
parallel lives	The approach in SSADM V4 for modelling asynchronous cycles in a single ELH, suitable for modelling within an application. This guide recommends using distinct aspects with separate ELHs, for three reasons: aspects address more general concepts of parallel behaviour, of which the parallel ELH within an application is a special case; aspects provide possibilities for partitioning within an application, to allow analysis and Logical Design to proceed in parallel; there are some situations that cannot be modelled with parallel lives.
PDI	See Process-Data Interface.
process-data interface	Documents how the Logical Data Model can be mapped on to the Physical Data Design, showing how it interfaces with the Physical Processing Specification. The PDI allows the designer to implement the logical update and enquiry processes as physical programs, independently of the physical database structure.
RDA	Abbreviation for Relational Data Analysis
Requirements Analysis	The objective is to produce the Analysis of Requirements. Within this the Selected Business System Option will define the scope of further investigation. This module has two stages: • Stage 1 Investigation of Current Environment • Stage 2 Business System Options.

Requirements Catalogue	The central repository for information covering all identified requirements, both functional and non-functional. Each entry is textual and describes a required facility or feature of the proposed system.
Smalltalk	A programming language embodying the major object-oriented concepts.
SSADM Rationale	A definition of the essential characteristics of SSADM, described in the ISE Library volume *Customising SSADM*
Soft Systems Methodology	A methodology directed at modelling of Human Activity Systems (see *Systems Thinking, Systems Practice* by Checkland P.B. and published in 1981 by John Wiley and Sons), which may be mapped on to organisation structures to describe business systems, and which often need support from IT systems. CCTA's ISE Library volume *Applying Soft Systems Methodology to an SSADM Feasibility Study* describes how Soft Systems Methodology can be used in an SSADM Feasibility Study.
persistent data	Data that needs to be stored beyond the lifetime of a single program execution.
SSM	See Soft Systems Methodology.
sub-type entity	An entity in a class hierarchy that in addition to its own properties is assumed to have all the properties of the super-types in the line of super-types above it in the class hierarchy.
sub-type node	A node in a class hierarchy at which there is a sub-type entity.
super-entity	Synonym for super-type entity.
super-event	A super-event occurs where two or more event-effects share the same operation set and end in the same entity state.

A common process identified as a selection component in an ELH, invoked by each of the events in the selection. The super-event may be propagated through other ELHs instead of the events that select it; this leads to simpler ELHs and a common UPM. |

super-type entity	An entity which has all the properties of entities lower down the class hierarchy.
super-type node	A node in a class hierarchy where the common properties of entities at nodes lower down the class hierarchy are recorded.
tailoring reuse	Means adapting an old component to make a new one. This is indirect reuse. It is more widespread but much less effective than direct reuse. Direct reuse means using a component without amending it: the hope is that the component can be implemented without any design or testing effort, because it is already tried and tested.
true reuse	Means extending the use of an existing component, not taking a copy of the component or creating a new component. Extending a component is only possible within the scope of one system. The main emphasis of the volume is on true reuse within a system, or a logical system which links several physical systems.
TNF	An abbreviation for third normal form. It describes the state of data after the application of the first three relational data analysis techniques.
TNF relation	A relation that is in third normal form. Relations may exist in fourth or higher normal forms.
undisciplined quit	A technique for modelling recognition problems in ELHs. It is less formal than the alternative, the disciplined quit. The major differences from the undisciplined quit are: *Quit/Resume* means 'the event labelled with R(*esume*) may occur **after** the event labelled with Q(*uit*)'; the scope of the recognition problem is not defined (*Quits* and *Resumes* may be placed wherever they are needed). The approach can provide good descriptions of recognition problems and their solutions but not formal specifications, because Quits and Resumes can over-ride the structure of the ELH. This means that some ELHs cannot be validated automatically and possibilities for automated derivation of ECDs and UPMs are limited.
Update Process Model	Is a structure diagram for update (event) processing and the associated operations list. The UPM is based on the

Entity Life Histories, which provide a data-oriented view of the system, and the associated Effect Correspondence Diagrams, which provide an event-oriented or process-oriented view of the system.

UPM See Update Process Model.

user role A user role is defined as a collection of job holders who share a large proportion of common tasks for which they may need support from an IT system.

Index

3-schema Specification Architecture 6, 8-10, 20, 21, 112, 126, 127, 135-137, 159, 160, 162, 163
   conceptual model 5, 19, 21, 23, 24, 27, 28, 31, 112, 124, 125, 127, 129, 131, 136, 159, 160, 162, 163
   conceptual schema 160
   external design 6, 21, 23, 27, 111, 115, 117, 125-127, 136, 159, 162
   external schema 162
   internal design 6, 10, 21, 23, 111, 119, 125, 127, 136, 159, 163
   internal schema 163
Aggregate 5, 19, 27, 29, 57, 155
Aggregation 27, 29, 37, 45, 46, 56, 57, 72, 128-130, 134, 159
Application 5, 9, 11, 15, 16, 29, 32, 43, 44, 46, 47, 57, 80, 129, 146, 157, 159, 165, 167
   application partitioning 9, 16, 29, 47, 129, 146, 157
Aspect 29, 30, 43-45, 51-56, 99, 101, 102, 104, 129, 151, 159
   aspects and aggregation 45
   aspects and inheritance 45
   basic existence 44, 45
Attribute 29, 44, 55, 56, 62, 68, 79, 80, 106, 115, 140, 144, 151, 152, 155, 159, 163-165
Batch 116, 121, 162
   batch input/output 162
Business 2, 3, 26, 39, 42, 43, 49, 50, 64-66, 68, 109, 133, 136, 159, 160, 162, 165, 166
Business activities 159, 160
Business activity 162
Business System Option 165
Class 5, 6, 11, 12, 14, 16, 19, 25, 30, 32, 33, 35-37, 42, 43, 45, 47, 49, 50, 52, 53, 55-59, 61, 63, 65-68, 71, 72, 87, 94-96, 101-104, 106, 107, 106, 109, 111, 113, 114, 116, 119, 120, 123, 128-134, 155, 159, 163, 165-167
Class hierarchy 14, 30, 32, 35, 37, 42, 43, 47, 53, 56-58, 61, 67, 71, 72, 96, 128-131, 134, 155, 159, 163, 166, 167
Client/server 24, 124, 126
Code 19, 53, 58, 82, 121, 127, 129-131, 160, 164

Communication 117, 119, 127, 160
Conceptual model 5, 19, 21, 23, 24, 27, 28, 31, 112, 124, 125, 127, 129, 131, 136, 159, 160, 162, 163
Conceptual schema 160
Corporate entity 45
Correspondence 8, 23, 28, 71, 73, 75, 79, 104, 106, 107, 109, 110, 117, 126, 134, 155, 161, 168
 one-to-one correspondence 79
Data 3, 5, 7-9, 11-13, 16, 17, 20, 21, 23, 26-32, 35-37, 43, 45, 46-52, 54, 55, 58-60, 63-68, 71, 92, 95, 108, 111, 112, 113-116, 118, 119, 122, 127, 128, 130, 133, 134, 136, 139, 141-145, 153, 160-168
 data storage 163
 stored data 29, 163
Data flow model 160, 161
Data modelling 5, 9, 17, 26-30, 35-37, 47, 68, 71
Database 6-10, 17, 21, 23, 25, 28, 29, 32, 37, 53, 63, 67, 68, 91, 108, 122, 128-131, 134, 136, 139, 152, 161, 163, 164, 165
Database management system 17, 25, 161
DBMS 131, 161
DFM 161
Dialogue 55, 126, 136
Dialogue design 136
Direct invocation 6, 104, 109, 112, 113, 117-119, 121, 122, 123
Divide 57, 81
Dividing 146
EAP 73, 161
Effect correspondence diagram 73, 155, 161
 ECD 27, 28, 53, 72-74, 83, 84, 87, 88, 90, 92, 94, 100, 101, 104, 109, 110, 112, 113, 116, 117, 119, 121, 122, 134, 139, 161
Encapsulation 5, 6, 11-13, 15, 19, 25, 53, 72, 101, 108, 162
Enquiry 6, 15, 21, 23, 31, 45, 72, 73, 80, 106, 107, 126, 131, 160-163, 165
Enquiry access path 73, 161
Enquiry process model 161, 162, 165
 EPM 162
Entity 5, 6, 17, 20, 21, 23, 25, 26, 28, 31-33, 35-37, 40, 42, 43-52, 54, 56, 57, 59, 61-63, 65-69, 71-76, 79-82, 88, 90, 92, 94, 95, 101, 102, 104, 112, 113, 118, 119,

123, 127, 133, 134, 139-143, 146, 151, 152, 155, 159, 161, 162-168
different behaviour 46
Entity aspect 45, 101, 102, 104
Entity life histories 23, 25, 75, 168
Entity life history 65, 155, 161, 162
   ELH 25, 32, 42, 52, 53, 75-77, 79, 80, 82, 83, 85, 86, 89, 90, 94, 95, 97, 98, 97-99, 102, 147, 148, 150-152, 161, 165-167
   entity life history analysis 65
   parallel life 98
   parallel lives 44, 97, 134, 165
Entity-event model 42, 52, 54, 80, 134, 139, 151
Entity-event modelling 5, 6, 20, 32, 35, 36, 49, 52, 56, 68, 69, 151
   event 5, 6, 15, 20, 23, 25-29, 32, 33, 35, 36, 42, 43, 45, 49, 52, 54, 56, 65, 68, 69, 71-77, 79-85, 87-90, 92, 94, 95, 99, 100, 102, 104-107, 109, 111-121, 123, 124, 126, 127, 129, 131-134, 139-146, 151-153, 155, 160, 161, 162, 163, 166-168
   event identification 81
   super-event 27, 28, 32, 72, 82-85, 87-90, 94, 99, 100, 104, 105, 109, 111, 118-120, 134, 153, 166
Event manager 23, 106, 109, 112, 113, 115-117, 119-121, 124, 131
External design 6, 21, 23, 27, 111, 115, 117, 125-127, 135, 136, 159, 162
External output 111, 112, 115, 116, 122
External schema 162
Feasibility Study 166
Function 159-162
Graphical user interface 8, 9, 162
   menus 162
GUI 8, 9, 17, 23, 27, 125, 157, 162
IM Library Volume
   Data Management 3
Implementation 5-10, 13, 14, 16, 19, 23, 24, 30, 37, 53, 80, 91, 92, 106, 108, 109, 112, 121, 126, 128, 129, 134, 136, 139, 152, 159, 160, 163, 164
Incremental delivery 16, 163
Inheritance 5, 6, 11, 14, 37, 40, 41, 45, 52, 56, 95, 100, 125, 126, 155, 159, 163
Integrating 102
Integration 9, 16, 29, 47, 129, 146, 157

Internal design 6, 10, 21, 23, 111, 119, 125, 127, 136, 159, 163
Internal schema 163
Investigation of current environment 165
Invocation message 111, 113, 114, 117, 132
ISE Library
  Application Partitioning and Integration with SSADM 9, 16, 29, 47, 129, 146, 157
ISE Library Volume 8, 16, 23, 47, 129, 135, 146, 166
  An Introduction to Reuse 17, 157
    Applying Soft Systems Methodology to an SSADM Feasibility Stud 166
  Customising SSADM 8, 9, 135, 157, 166
  Distributed Systems: Application Development 29, 157
  Managing Reuse 17, 157
IT System 159, 168
LDM 5, 25-29, 31, 32, 35-37, 41-43, 48, 49, 52-57, 59, 63, 64, 67-69, 71-73, 95, 96, 98, 100, 128, 133, 134, 139, 146, 151, 159, 160, 163
Logical data model 5, 8, 23, 27, 28, 35, 36, 48, 59, 65, 95, 161, 163-165
Logical design 161, 165
Merging 56, 102
Message 5-7, 9, 11-13, 15, 19, 23, 25, 27, 72, 92, 101, 102, 106, 109-118, 121-123, 131, 132, 134, 153, 164
Message type 102, 111, 114, 132
Message-passing 5-7, 9, 11, 15, 19, 23, 27, 72, 106, 109, 110, 112, 113, 118, 121, 122, 134, 164
Method 3, 6, 7, 9, 12, 15, 19-21, 23, 25-29, 52, 73-76, 79, 80, 82, 101-104, 106, 107, 109-111, 114, 119-121, 123, 125, 155, 160, 164
Method invocation 6, 9, 109
Network invocation 6, 109, 112, 113, 123
Object 1, 5, 7-15, 17, 19, 20, 23, 25-29, 35, 36, 49, 51, 52, 57, 63, 64, 66, 68, 72, 80, 101-104, 106, 108-121, 123, 124, 125-127, 129-132, 134, 155, 157-160, 162-166
Operation 6, 10, 25, 28, 82, 83, 90, 92, 94, 95, 101, 106, 134, 151-153, 155, 166
Operations list 161, 165, 167
Optimisation 6, 48, 71, 76, 77, 82, 121
Optimise 77, 121, 134
Outputs 23
Parallel lives 44, 97, 134, 165

Index

Partition 123, 136, 146
Partitioned 9, 16, 129
Partitioning 9, 16, 29, 46, 47, 129, 146, 157, 165
PDI 129, 131, 132, 134, 136, 163, 165
Physical design 52, 108
Process 6, 9, 11-13, 17, 21, 23, 25, 27, 28, 32, 53, 71-73, 75, 80, 90, 91, 99, 100, 102, 104, 106, 109, 121, 127, 135, 136, 160-163, 165-168
Process-data interface 165
Program-data interface 21
Programme 3, 17
Prototype 126
RDA 52, 53, 55, 56, 133, 165
Real-world 28, 43, 44, 46, 64, 65, 159
Relational Data Analysis 52, 54, 165, 167
  third normal form 27, 167
    TNF 133, 167
Relational database 9, 10, 29, 131, 152
Relationship 21, 39, 43, 44, 62, 79, 146, 151, 152, 155, 164
Requirements Analysis 165
Response data 111, 115, 118, 119, 122
Rolling down 131
Rolling up 37, 56
Scope 127, 160, 161, 163, 165, 167
Sealed packet 114-116, 119, 122
Selective locking 111, 131
Shared server 129, 131
Soft systems methodology 166
SSADM 1-3, 5-10, 16, 17, 19-30, 36, 47, 49-51, 55, 71-73, 75-79, 82, 92, 102, 121, 125, 127, 129, 133, 135, 136, 146, 151, 152, 155, 157-160, 165, 166
SSADM Rationale 127, 166
SSADM V4 Reference Manuals 27, 55, 75, 82, 102, 121, 157
Sub-type 14, 30, 38, 45, 53, 54, 56, 57, 61, 67, 72, 98, 100, 130, 131, 133, 161, 163, 166
Subsystem 45
Super-event 27, 28, 32, 72, 82-85, 87-90, 94, 99, 100, 104, 105, 109, 111, 118-120, 134, 153, 166
Super-type 14, 30-32, 36-38, 41, 42, 53, 56, 58, 61, 62, 65, 66-68, 72, 128-131, 161, 166, 167
Three schemata 127
Two-phase commit 23, 117, 122

Universal function model 159
Update 8, 15, 21, 23, 25, 60, 71, 75, 79, 80, 91, 119, 121, 129, 132, 134, 140, 143, 144, 160-165, 167, 168
Update process model 121, 165, 167, 168
Updated 15
Updating 63, 79, 152
UPM 92, 121, 139, 153, 166-168
User role 168
Workstation 126, 127
'passed-through' data 113, 115